CONTEMPORARY'S
REAL NUMBERS
Developing Thinking Skills in Math
Estimation 1: Whole Numbers and Decimals

Allan D. Suter

Project Editor
Kathy Osmus

CONTEMPORARY BOOKS

a division of NTC/CONTEMPORARY PUBLISHING GROUP
Lincolnwood, Illinois USA

ISBN: 0-8092-4214-1

Published by Contemporary Books,
a division of NTC/Contemporary Publishing Group, Inc.,
4255 West Touhy Avenue,
Lincolnwood (Chicago), Illinois, 60712-1975 U.S.A.
© 1991 by Allan D. Suter
All rights reserved. No part of this book may be reproduced,
stored in a retrieval system, or transmitted in any form or by any means,
electronic, mechanical, photocopying, recording or otherwise,
without prior permission of the publisher.
Manufactured in the United States of America.

0 1 2 3 4 5 6 7 8 9   C(K) 16 15 14 13 12 11 10

Editorial Director	*Cover Design*
Caren Van Slyke	Lois Koehler
Editorial	*Illustrator*
Lisa Black	Ophelia M. Chambliss-Jones
Janice Bryant	
Karin Evans	*Art & Production*
Ellen Frechette	Princess Louise El
Lynn McEwan	
Steve Miller	*Typography*
Karen Schenkenfelder	Design/Americom
Seija Suter	Chicago, Illinois

Editorial Production Manager
Norma Fioretti

Production Editor
Jean Farley Brown

Production Assistant
Marina Micari

Cover photo by Michael Slaughter

CONTENTS

Learning About Estimation

Estimation helps you find an amount that is close to the exact answer. Estimation is useful when checking the accuracy of answers and often more practical than finding the exact answer.

Listed below are exact numbers paired with their approximate values (values close to the original numbers). Notice the key words in **bold** type, which show the values aren't exact. Approximate values are often used because they are easier and quicker to work with.

Exact Numbers	Approximate Values
18 students	**About** 20 students
$8,994	**Nearly** $9,000
188 miles	**Almost** 200 miles
9,850 years ago	**Around** 10,000 years ago
Final tally: 1,090	**Close to** 1,000
616,722 tourists	A **little more** than 600,000 tourists

▶ Answer the following questions using the information above.

1. Are all of the approximate values close to the exact numbers? _____

2. Why do you think we use approximate numbers? _____

   _____

3. What are some key words that are used in estimation? _____

   _____

▶ Circle the answer that makes the most sense.

4. The cost of 2 hamburgers, french fries, and a soft drink at a fast food restaurant would be about

   a) $15.   **b)** $50.   **c)** $5.

5. The cost of a new 19-inch television would be around

   a) $40.   **b)** $400.   **c)** $4,000.

6. Ken worked 20 hours and earned nearly

   a) $200.   **b)** $2,000.   **c)** $20.

Are the Answers Reasonable?

Estimation does not require you to find the amount with pencil and paper.

▶ Complete the following statements by writing in a reasonable estimate.

1. I usually sleep about _____ hours each night.

2. It takes me about _____ minutes to wash the dishes.

3. I will spend nearly _____ hours on homework this week.

4. My birthday is about _____ months away.

5. There are about _____ students in my math class.

6. I plan to save around $_____ next month.

▶ Use "number sense" (what you know about numbers) to tell whether each statement is reasonable or not. Circle the letter next to your answer.

7. A medium-sized car gets about 8 miles per gallon.   **a)** reasonable   **b)** not reasonable

8. Leslie walked a mile in 20 minutes.   **a)** reasonable   **b)** not reasonable

9. Molly drove 250 miles in 2 hours.   **a)** reasonable   **b)** not reasonable

10. Bert bought 2 pencils for 30 cents.   **a)** reasonable   **b)** not reasonable

11. The average-sized man weighs '300 pounds.   **a)** reasonable   **b)** not reasonable

12. One-half gallon of milk will fill less than 2 glasses.   **a)** reasonable   **b)** not reasonable

Decide When to Estimate

In some cases an exact answer is not needed.

▶ Read each situation below and decide whether it makes more sense to estimate or to find the exact answer. Circle your answer.

1. How much time does it take to paint a house?

    a) estimate   **b)** exact

2. How much paint is needed to paint a house?

    a) estimate   **b)** exact

3. What is the balance in Mrs. Holt's checking account?

    a) estimate   **b)** exact

4. Will there be enough money to pay for the groceries?

    a) estimate   **b)** exact

5. How much change did you get back after paying for the groceries?

    a) estimate   **b)** exact

6. What was the population increase for the state of Florida?

    a) estimate   **b)** exact

7. What time is the airplane scheduled to leave the airport?

    a) estimate   **b)** exact

Estimating Metric Measurements

The metric system is used throughout the world to measure items. Below are examples that compare different metric measurements to things we know.

The width of a paper clip is almost 1 centimeter.

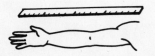 The length of a man's arm is about 1 meter.

It takes about 10 minutes to walk 1 kilometer (a little farther than $\frac{1}{2}$ of a mile).

▶ Use the examples above to help you choose the best estimate for each of the following measurements. Circle your answers.

1. The distance from Chicago to St. Louis is about 480

   a) centimeters.   **b)** meters.   **c)** kilometers.

2. The average height of a basketball player is about 2

   a) meters.   **b)** kilometers.   **c)** centimeters.

3. The length of a man's tennis shoe is about 30

   a) kilometers.   **b)** centimeters.   **c)** meters.

4. A carpenter used nails that were 6

   a) centimeters.   **b)** kilometers.   **c)** meters.

5. The Hearty Hustle Walk/Run Race is about 5

   a) meters.   **b)** centimeters.   **c)** kilometers.

6. The length of the living room was close to 10

   a) kilometers.   **b)** centimeters.   **c)** meters.

Front-End Estimation

Front-end estimation is a method for finding quick, easy, and reasonable answers in your head. When you use front-end estimation, add the digits that have the largest place values. These are the front-end digits.

Example A

To estimate, add the front-end digits 4, 2, and 3.

$$
\begin{array}{rcl}
47 &=& \textbf{4} \text{ tens} + 7 \text{ ones} \\
23 &=& \textbf{2} \text{ tens} + 3 \text{ ones} \\
+35 &=& \textbf{3} \text{ tens} + 5 \text{ ones} \\
\hline
&=& \textbf{9} \text{ tens} \\
&=& 90 \longleftarrow \text{ estimate}
\end{array}
$$

Example B

To estimate, add the front-end digits 3, 1, and 2.

$$
\begin{array}{rcl}
357 &=& \textbf{3} \text{ hundreds} + 5 \text{ tens} + 7 \text{ ones} \\
126 &=& \textbf{1} \text{ hundred} + 2 \text{ tens} + 6 \text{ ones} \\
+230 &=& \textbf{2} \text{ hundreds} + 3 \text{ tens} + 0 \text{ ones} \\
\hline
&=& \textbf{6} \text{ hundreds} \\
&=& 600 \longleftarrow \text{ estimate}
\end{array}
$$

Front-end estimation will always give you an estimate that is less than the exact answer. (This is called an **underestimate**.)

▶ Estimate the following answers using only the front-end digits.

1.
$$
\begin{array}{r}
27 \\
56 \\
+23 \\
\hline
\end{array}
$$
__9__ tens = __ __ estimate

4.
$$
\begin{array}{r}
542 \\
254 \\
+118 \\
\hline
\end{array}
$$
__ hundreds = __ __ __ estimate

2.
$$
\begin{array}{r}
34 \\
22 \\
+16 \\
\hline
\end{array}
$$
__ tens = __ __ estimate

5.
$$
\begin{array}{r}
450 \\
702 \\
+628 \\
\hline
\end{array}
$$
__ __ hundreds = __ , __ __ __ estimate

3.
$$
\begin{array}{r}
12 \\
46 \\
+28 \\
\hline
\end{array}
$$
__ tens = __ __ estimate

6.
$$
\begin{array}{r}
325 \\
719 \\
+567 \\
\hline
\end{array}
$$
__ __ hundreds = __ , __ __ __ estimate

Estimate Horizontal Forms

Numbers to be added are sometimes written in a row (horizontally). Front-end estimation is an easy way to get an approximate answer when numbers are written in this form.

Example

The digits 6, 2, and 7, which are at the beginning of the numbers, are called front-end digits. If you add the 6 tens, 2 tens, and 7 tens, you get 15 tens, or 150. This is a reasonable answer using front-end estimation.

6 tens   2 tens   7 tens
↑   ↑   ↑
6 2 + **2** 6 + **7** 8 is about 1 5 0.

▶ Add the front-end digits to find each estimate.

3 tens   2 tens   9 tens
↑   ↑   ↑

1. **32** + **26** + **91** is about ___14___ tens or _____ .

2. **58** + **24** + **81** + **32** is about _____ .

3. **55** + **63** + **32** + **44** is about _____ .

4 hundreds   8 hundreds   1 hundred
↑   ↑   ↑

4. **424** + **869** + **108** is about _____ hundreds or _____ .

5. **515** + **347** + **752** + **105** is about _____ .

6. **632** + **413** + **930** + **210** is about _____ .

7 thousands   2 thousands   5 thousands
↑   ↑   ↑

7. **7,204** + **2,158** + **5,623** is about _____ thousands or _____ .

8. **8,065** + **6,781** + **3,436** + **4,106** is about _____ .

9. **2,881** + **3,016** + **4,224** is about _____ .

Different Place Values for Front-End Digits

Sometimes when we estimate, the front-end digits have different place values. With front-end estimation, be careful to add only the front-end digits with the largest place value.

Example	Step 1	Step 2
To estimate:	Look for the front-end digits with the largest place value.	Add the front-end digits, and estimate.

4,275	Think	**4,275**	**4,275**
382	zero. →	0 382	382
95	↘	0 095	95
+ **7,236**		+ **7,236**	+ **7,236**
			11 thousands = 11,000

▶ Add the front-end digits with the largest place value, then estimate.

1.
$$\begin{array}{r} \mathbf{2}\,34 \\ 25 \\ \mathbf{4}\,09 \\ +\quad 8 \\ \hline \end{array}$$

_6_ hundreds = __ __ __
estimate

2.
$$\begin{array}{r} \mathbf{6},249 \\ \mathbf{3},026 \\ 317 \\ +\quad 15 \\ \hline \end{array}$$

__ thousands = __ , __ __ __
estimate

3.
$$\begin{array}{r} \mathbf{5},135 \\ 692 \\ \mathbf{3},415 \\ 15 \\ +\mathbf{6},918 \\ \hline \end{array}$$

__ __ thousands = __ __ , __ __ __
estimate

4. 5 hundreds   3 hundreds
   ↙   ↙

560 + 37 + **345**

Front-end estimate: _____

5. 6,213 + 182 + 8,468

Front-end estimate: _____

6. 58 + 229 + 952 + 8

Front-end estimate: _____

7. 86 + 9 + 92 + 38

Front-end estimate: _____

Applications for Front-End Estimation

Estimation only requires a close, or approximate, answer.

▶ Use front-end estimation to solve the problems below.

1. About how many calories were in the food that was eaten for

   a) breakfast and lunch? _____

   b) breakfast, lunch, dinner, and snack? _____

Calories Eaten	
Breakfast	535
Lunch	820
Dinner	940
Snack	350

2. About how many miles were driven on

   a) days 1 and 2? _____

   b) days 1 through 4? _____

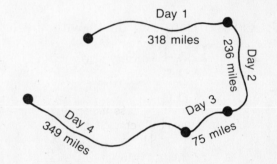

3. About how many people attended the fair on

   a) Monday and Tuesday? _____

   b) Thursday and Friday? _____

   c) Monday through Friday? _____

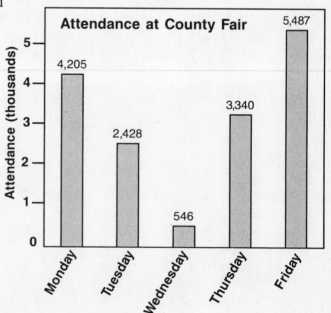

Greatest Place Value

When estimating in real life, we often round numbers to the greatest place value. That place value is called the **lead digit** (the first digit on the left). To round a number to the lead digit, look at the next greatest place value to the right.

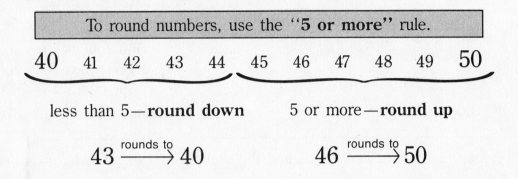

less than 5—**round down**   5 or more—**round up**

$$43 \xrightarrow{\text{rounds to}} 40 \qquad\qquad 46 \xrightarrow{\text{rounds to}} 50$$

If the digit to the right of the lead digit is less than 5 (4, 3, 2, 1, 0), change all the numbers to 0 except the lead digit.

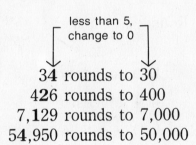

34 rounds to 30
426 rounds to 400
7,129 rounds to 7,000
54,950 rounds to 50,000

If the digit to the right of the lead digit is 5 or more (5, 6, 7, 8, 9), add 1 to the lead digit. Then change all the other digits to 0.

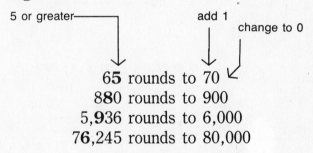

65 rounds to 70
880 rounds to 900
5,936 rounds to 6,000
76,245 rounds to 80,000

▶ Based on the examples shown above, circle the correct rounded number from the 2 choices.

1. 492 rounds to 400 or ⑤⓪⓪

2. 33 rounds to 30 or 40

3. 6,570 rounds to 6,000 or 7,000

4. 2,487 rounds to 2,000 or 3,000

5. 29,301 rounds to 20,000 or 30,000

6. 85 rounds to 80 or 90

7. 17,254 rounds to 10,000 or 20,000

▶ Round each number to the lead digit.

8. 429 rounds to _____

9. 3,086 rounds to _____

10. 58,697 rounds to _____

11. 98 rounds to _____

12. 537 rounds to _____

13. 1,569 rounds to _____

14. 73,850 rounds to _____

Rounding to Estimate

When adding, a quick way to estimate is to round the numbers to their lead digits, then add.

Example A

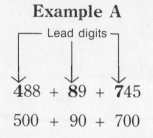

Lead digits

$\overset{\downarrow}{4}88 + \overset{\downarrow}{8}9 + \overset{\downarrow}{7}45$

$500 + 90 + 700$

Rounded estimate: 1,290

Example B

The lead digits are 5, 2, and 6.

$$
\begin{array}{rl}
\mathbf{5}65 \text{ rounds to} & 600 \\
\mathbf{2}14 \text{ rounds to} & 200 \\
+\ \mathbf{6}9 \text{ rounds to} & +\ 70 \\
\hline
\text{Rounded estimate:} & 870
\end{array}
$$

▶ Rewrite the following problems by rounding the numbers to their lead digits, then add to find the rounded estimate.

1. $625 + 980 + 45$

   $\underline{600} + \underline{1,000} + \underline{50}$

   Rounded estimate: _____

2. $63 + 737 + 109$

   $\underline{80} + \underline{800} + \underline{100}$

   Rounded estimate: _____

3. $37 + 315 + 56$

   $\underline{40} + \underline{300} + \underline{60}$

   Rounded estimate: _____

4. $893 + 74 + 239$

   $\underline{800} + \underline{70} + \underline{200}$

   Rounded estimate: _____

5. 126 rounds to __100__

   242 rounds to __200__

   $+\ 97$ rounds to __100__

   Rounded estimate: __400__

6. 965 rounds to __1000__

   44 rounds to __40__

   $+160$ rounds to __200__

   Rounded estimate: _____

7. 539 rounds to __500__

   82 rounds to _____

   $+\ 75$ rounds to _____

   Rounded estimate: _____

8. 46 rounds to _____

   309 rounds to _____

   $+\ 24$ rounds to _____

   Rounded estimate: _____

Rounding Larger Numbers to Add

Rounding numbers to their lead digits can be a great help when estimating.

Example A

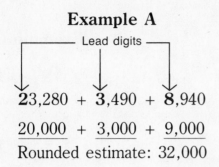

Lead digits

23,280 + 3,490 + 8,940

20,000 + 3,000 + 9,000

Rounded estimate: 32,000

Example B

The lead digits are 7, 2, and 4.

78,356 rounds to   80,000

2,083 rounds to   2,000

+45,614 rounds to +50,000

Rounded estimate: 132,000

▶ Rewrite the following problems by rounding the numbers to their lead digits, then add to find the rounded estimate.

1. 46,735 + 73,142 + 8,914

   50,000 + 70,000 + 9,000

   Rounded estimate: _____

2. 22,372 + 8,281 + 370

   _____ + _____ + _____

   Rounded estimate: _____

3. 7,590 + 49,240 + 1,736

   _____ + _____ + _____

   Rounded estimate: _____

4. 68,512 + 15,126

   _____ + _____

   Rounded estimate: _____

5.   57,239 rounds to _____

       6,170 rounds to _____

   + 90,826 rounds to _____

   Rounded estimate: _____

6.   21,386 rounds to _____

       3,499 rounds to _____

   + 88,702 rounds to _____

   Rounded estimate: _____

7.   97,394 rounds to _____

      41,508 rounds to _____

   + 10,800 rounds to _____

   Rounded estimate: _____

8.   26,317 rounds to _____

       5,086 rounds to _____

   + 3,728 rounds to _____

   Rounded estimate: _____

Grouping to 100

Grouping numbers makes it easy to estimate. Look for numbers that go together to make about 100.

Example A

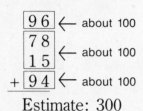

96 ← about 100
78 ← about 100
15
+ 94 ← about 100
Estimate: 300

Example B

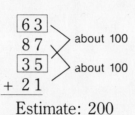

63 } about 100
87 }
35 } about 100
+ 21 }
Estimate: 200

Example C

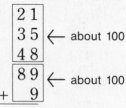

21
35 ← about 100
48
89 ← about 100
+ 9
Estimate: 200

▶ Estimate the following sums by looking for groups of numbers that make about 100.

1.
74 ← about 100
23
81 ← about 100
+22
Estimate:_____

4.
14
83
51
+47
Estimate:_____

7.
98
35
49
 8
+96
Estimate:_____

2.
97
19
98
+73
Estimate:_____

5.
38
95
58
+99
Estimate:_____

8.
98
72
93
25
+95
Estimate:_____

3.
45
88
13
+49
Estimate:_____

6.
18
38
60
+73
Estimate:_____

Adjusting Front-End Estimation

- Front-end estimates are often adjusted up to get closer to the exact answer.

- When you estimate, all reasonable answers are acceptable.

Example

Step 1

Add the front-end digits.

$$
\begin{array}{r}
\mathbf{4}\,6\,9 \\
\mathbf{8}\,4\,2 \\
\mathbf{7}\,8\,9 \\
+\,\mathbf{5}\,1\,5 \\
\hline
\end{array}
$$

24 hundreds
= 2,400

Step 2

Group the digits that are not used for the front-end estimation.

$$
\begin{array}{r}
4\,\boxed{6\,9} \leftarrow \text{about 100}\\
8\,\boxed{4\,2}\\
7\,\boxed{8\,9} \leftarrow \text{about 100}\\
+\,5\,\boxed{1\,5}\\
\hline
\end{array}
$$

About 200 more

Step 3

Add the sums found in steps 1 and 2 to find the adjusted estimate.

Front-end sum → 2,400
+   200 more

Adjusted estimate: 2,600

▶ Find the front-end estimate. Then use grouping to adjust the front-end estimate.

1.
$$
\begin{array}{r}
\mathbf{3}\,\boxed{8\,3} \\
\mathbf{5}\,5\,8 \\
\mathbf{2}\,\boxed{1\,4} \\
+\,\mathbf{6}\,3\,5 \\
\end{array}
$$
about 100
about 100

a) 1,600 and about _____ more

b) Adjusted estimate:_____

2.
$$
\begin{array}{r}
\mathbf{4}\,\boxed{9\,6} \leftarrow \text{about } \_\_\_\_ \\
\mathbf{5}\,\boxed{4\,5} \\
\mathbf{3}\,\boxed{3\,9} \leftarrow \text{about } \_\_\_\_ \\
+\,\mathbf{1}\,\boxed{0\,8} \\
\end{array}
$$

a) _____ and about _____ more

b) Adjusted estimate:_____

3.
$$
\begin{array}{r}
\mathbf{4}\,\boxed{1\,9} \\
\mathbf{3}\,5 \\
\mathbf{3}\,6\,0 \\
+\,\boxed{7\,2} \\
\end{array}
$$
about _____   about _____

a) _____ and about _____ more

b) Adjusted estimate:_____

4.
$$
\begin{array}{r}
\mathbf{3}\,3\,2 \\
\boxed{9\,8} \leftarrow \text{about } \_\_\_\_ \\
\boxed{9\,6} \leftarrow \text{about } \_\_\_\_ \\
+\,\mathbf{8}\,7\,0 \\
\end{array}
$$
about _____

a) _____ and about _____ more

b) Adjusted estimate:_____

Adjusting Larger Numbers

Adjust the sum of the front-end digits by grouping the unused digits to 1,000. Adjusting the front-end digits will get closer to the exact answer.

Example

Step 1

Add the front-end digits.

```
  3 , 3 9 7
  1 , 7 3 0
  6 , 5 9 3
  2 , 3 4 2
+ 4 , 9 7 5
```
  16 thousands
  = 16,000

Step 2

Group the digits that are not used for your front-end estimation.

```
  3 , 3 9 7
  1 , 7 3 0    about 1,000
  6 , 5 9 3
  2 , 3 4 2    about 1,000
+ 4 , 9 7 5  ← about 1,000
```
About 3,000 more

Step 3

Add the sums found in steps 1 and 2 to find the adjusted estimate.

Front-end sum → 16,000
+ 3,000 more

Adjusted estimate: 19,000

▶ Find the front-end estimate. Then use grouping to adjust the front-end estimate.

1.
```
  4 , 964  ← about _____
  6 , 482
  5 , 825  > about _____
  2 , 509  > about _____
+ 8 , 247
```
a) __ ,000 and about _____ more

b) Adjusted estimate: _____

3.
```
  5 , 123
  4 , 680
  7 , 219
  5 , 563
+ 9 , 408
```
a) __ ,000 and about _____ more

b) Adjusted estimate: _____

2.
```
  3 , 132
  5 , 885
  6 , 990
  1 , 456
+ 2 , 518
```
a) __ ,000 and about _____ more

b) Adjusted estimate: _____

4.
```
  7 , 999
      793
  2 , 106
  4 , 892
+     225
```
a) __ ,000 and about _____ more

b) Adjusted estimate: _____

Apply Your Skills

▶ Use front-end, grouping, or rounding strategies to estimate. Any reasonable answer is acceptable.

1.

Week	Wages
1	$345
2	$395
3	$270

a) Estimate the total wages. _____

b) Do you think your estimate is greater or less than the exact answer? _____

c) Find the exact answer to see how close your estimate is. _____

2. 2,350 + 3,244 + 6,821

a) Estimate: _____

b) Do you think your estimate is greater or less than the exact answer? _____

c) Find the exact answer to see how close your estimate is. _____

3.

$$
\begin{array}{r}
97 \\
28 \\
71 \\
19 \\
+\,83 \\
\hline
\end{array}
$$

a) Estimate: _____

b) Do you think your estimate is greater or less than the exact answer? _____

c) Find the exact answer to see how close your estimate is. _____

4.

Desk	$405
Chair	$74
Folding table	$45
Stacking chairs	$98

a) Estimate the total cost of the items listed above. _____

b) Do you think your estimate is greater or less than the exact answer? _____

c) Find the exact answer to see how close your estimate is. _____

5. 12,482 + 2,723 + 970

a) Estimate: _____

b) Do you think your estimate is greater or less than the exact answer? _____

c) Find the exact answer to see how close your estimate is. _____

6.

$$
\begin{array}{r}
6,230 \\
629 \\
3,242 \\
2,607 \\
+\ \ 395 \\
\hline
\end{array}
$$

a) Estimate: _____

b) Do you think your estimate is greater or less than the exact answer? _____

c) Find the exact answer to see how close your estimate is. _____

Estimating to Subtract

Estimation is a skill you use every day. Circle the answer that makes the most sense.

1. Lori bought a sandwich, french fries, and a soft drink. She paid for the food with a $20 bill. About how much change will she get back?

   a) $5   **b)** $15   **c)** $2

2. About how much is the difference in price?

   a) $8,000   **b)** $20,000   **c)** $50,000

3. Oscar saved $800. About how much more money will he need to buy the new motorcycle?

   a) $20,000   **b)** $200   **c)** $5,500

Rounding to Subtract

Rounding numbers to subtract is a method for finding quick, easy, and reasonable answers in your head. Round each number to the greatest place value (lead digit), then estimate the difference.

Example A
887 rounds to   900
−344 rounds to −300
Rounded estimate: 600

Example B
9,082 rounds to   9,000
−4,775 rounds to −5,000
Rounded estimate: 4,000

▶ Round each number to the lead digit, then estimate the difference.

1.
71 rounds to   <u>70</u>
− 38 rounds to <u>− 40</u>
Rounded estimate: _____

5.
692
−478
Rounded estimate: _____

2.
807 rounds to _____
− 216 rounds to _____
Rounded estimate: _____

6.
5,671
− 1,084
Rounded estimate: _____

3.
4,930 rounds to _____
− 1,038 rounds to _____
Rounded estimate: _____

7.
88
− 29
Rounded estimate: _____

4.
67,592 rounds to _____
− 48,108 rounds to _____
Rounded estimate: _____

8.
71,501
− 23,894
Rounded estimate: _____

Subtracting with Front-End Estimation

To estimate the difference using front-end estimation:

• Subtract the front-end digits.

• Compare the remaining digits for a closer estimate.

Example A

Step 1

Subtract the front-end digits (5, 2)
to find the front-end estimate.

$$\begin{array}{r} 562 \\ -224 \\ \hline \end{array}$$

3 hundreds

Step 2

Compare the remaining numbers,
62 and 24.

$$\begin{array}{r} 5|62 \\ -2|24 \\ \hline \end{array}$$ ← The top number is larger than the bottom number, so the answer is **greater than** 300.

Greater than 300

Example B

Step 1

Subtract the front-end digits (9, 3)
to find the front-end estimate.

$$\begin{array}{r} 932 \\ -376 \\ \hline \end{array}$$

6 hundreds

Step 2

Compare the remaining numbers,
32 and 76.

$$\begin{array}{r} 9|32 \\ -3|76 \\ \hline \end{array}$$ ← The top number is smaller than the bottom number, so the answer is **less than** 600.

Less than 600

▶ Circle the answer that gives the closer estimate.

1. $\begin{array}{r} 847 \\ -205 \\ \hline \end{array}$

   a) greater than 600

   b) less than 600

3. $\begin{array}{r} 919 \\ -426 \\ \hline \end{array}$

   a) greater than 500

   b) less than 500

5. $\begin{array}{r} 364 \\ -191 \\ \hline \end{array}$

   a) greater than 200

   b) less than 200

2. $\begin{array}{r} 548 \\ -173 \\ \hline \end{array}$

   a) greater than 400

   b) less than 400

4. $\begin{array}{r} 736 \\ -521 \\ \hline \end{array}$

   a) greater than 200

   b) less than 200

6. $\begin{array}{r} 580 \\ -375 \\ \hline \end{array}$

   a) greater than 200

   b) less than 200

Adjusting Front-End Estimation

Sometimes when you subtract, you will want to get a closer estimate.

Example A

Step 1

Find the front-end estimate.

$$\begin{array}{r} 93,251 \\ -37,648 \\ \hline 60,000 \end{array}$$

Step 2

Working with the first two digits will give an adjusted estimate. Here you must regroup.

8 13
$$\begin{array}{r} 9\cancel{3},251 \\ -37,648 \\ \hline 56,000 \end{array}$$

Use zeros as placeholders.

Example B

Step 1

Find the front-end estimate.

$$\begin{array}{r} 86,475 \\ -72,692 \\ \hline 10,000 \end{array}$$

Step 2

To find the adjusted estimate, no regrouping is necessary here.

$$\begin{array}{r} 86,475 \\ -72,692 \\ \hline 14,000 \end{array}$$

Use zeros as placeholders.

▶ Find the front-end and adjusted estimates.

1.
$$\begin{array}{r} 65,758 \\ -26,946 \end{array}$$

   a) Front-end estimate: __40,000__

   b) Adjusted estimate: _____

2.
$$\begin{array}{r} 54,629 \\ -21,075 \end{array}$$

   a) Front-end estimate: _____

   b) Adjusted estimate: _____

3.
$$\begin{array}{r} 49,172 \\ -15,384 \end{array}$$

   a) Front-end estimate: _____

   b) Adjusted estimate: _____

4.
$$\begin{array}{r} 7,348 \\ -3,902 \end{array}$$

   a) Front-end estimate: _____

   b) Adjusted estimate: _____

Zero at Work

Sometimes when you subtract, the front-end digits have different place values.

Example A

Step 1
Find the front-end estimate.

Think zero.
$$\begin{array}{r} 4,608 \\ -\ 0\ 532 \\ \hline 4,000 \end{array}$$

Step 2
To find the adjusted estimate, no regrouping is necessary.

$$\begin{array}{r} 4,608 \\ -\quad 532 \\ \hline 4,100 \end{array}$$

Example B

Step 1
Find the front-end estimate.

Think zero.
$$\begin{array}{r} 7,304 \\ -\ 0\ 882 \\ \hline 7,000 \end{array}$$

Step 2
To find the adjusted estimate, you must regroup.

$$\begin{array}{r} {}^{6}\ {}^{13} \\ 7,\cancel{3}04 \\ -\quad 882 \\ \hline 6,500 \end{array}$$

▶ Find the front-end and adjusted estimates.

1.
$$\begin{array}{r} 6,721 \\ -\quad 805 \end{array}$$
a) Front-end estimate: _6,000_
b) Adjusted estimate: _____

4.
$$\begin{array}{r} 9,673 \\ -\quad 925 \end{array}$$
a) Front-end estimate: _____
b) Adjusted estimate: _____

2.
$$\begin{array}{r} 5,562 \\ -\quad 195 \end{array}$$
a) Front-end estimate: _____
b) Adjusted estimate: _____

5.
$$\begin{array}{r} 7,380 \\ -\quad 642 \end{array}$$
a) Front-end estimate: _____
b) Adjusted estimate: _____

3.
$$\begin{array}{r} 1,735 \\ -\quad 436 \end{array}$$
a) Front-end estimate: _____
b) Adjusted estimate: _____

6.
$$\begin{array}{r} 3,421 \\ -\quad 279 \end{array}$$
a) Front-end estimate: _____
b) Adjusted estimate: _____

Real-World Applications

▶ Use the estimation skills you learned earlier to answer the following questions.

1. About what is the difference in price? _____

$18,842

$9,432

2. About how much is saved by buying the color television on sale? _____

Regular price • $628
Sale price • $479

3. About how many more fans attended the Reds game? _____

Opening Day Attendance
Tigers - 49,362
Reds - 63,882

4. Estimate the population increase for Elk County based on the census report below. _____

Census Report	
1980	43,184
1990	97,206

Make a Reasonable Guess

▶ Estimate the answers to the following questions.

1. The *Daily Press* sold about how many more newspapers than:

   a) the *Herald News?* _____

   b) the *Tribune?* _____

Newspaper Sales	
Daily Press	92,891
Herald News	7,125
Tribune	50,876

2. About what is the combined sales of all 3 newpapers? _____

3. Wooddale has about how many more students enrolled than:

   a) Northwestern? _____

   b) Lakeland? _____

   c) Franklin? _____

School Enrollment	
Northwestern	1,250
Lakeland	682
Franklin	455
Wooddale	1,813

4. About how many students are enrolled in all 4 schools? _____

5. About how much more money did the Tigers collect than:

   a) the Broncos? _____

   b) the Bulldogs? _____

   c) the Giants? _____

Fund-Raising Drive	
Team	Amount
Broncos	$2,785
Bulldogs	520
Tigers	9,968
Giants	5,129

6. About how much money was collected altogether? _____

What Is Reasonable?

When you are multiplying, estimation can give you quick and easy answers that are close to the exact answers.

▶ Estimate the price of each item shown below. Then circle the answer that makes the most sense.

$ _____

1. About how much will 4 gallons of paint cost?

    a) $5   **b)** $500   **c)** $50

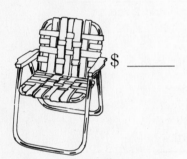

$ _____

2. About what is the cost of 4 lawn chairs?

    a) $40   **b)** $400   **c)** $1,200

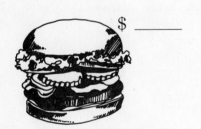

$ _____

3. About what is the cost of 6 hamburgers?

    a) $3   **b)** $18   **c)** $60

$ _____

4. About what is the cost of 3 new bicycles?

    a) $900   **b)** $90   **c)** $9,000

$ _____

5. About how much will 5 new tires cost?

    a) $1,600   **b)** $100   **c)** $500

Estimating with Multiplication

For a quick, reasonable estimate, round the factor that is greater than 10 to its lead digit (greatest place value). It is not necessary to round a one-digit factor.

Example

If Cari's house payments are $379 each month, about how much will she pay in 8 months?

Step 1
Write a number sentence for the problem.

$379 × 8

↑ Round up, since 7 is greater than 5.

Step 2
Round $379 to 400.

$400 × 8

Step 3
Multiply 4 × 8, and add 2 zeros.

$400 × 8 = $3,200

Add 2 zeros.

Cari's house payments are about $3,200 for 8 months.

▶ Round the digit greater than 10, then estimate.

1. 39 × 6
   40 × 6 = __ __ 0
   Add 1 zero.

2. 652 × 9
   700 × 9 = __ , __ 0 0
   Add 2 zeros.

3. 4,085 × 4
   4,000 × 4 = __ __ , 0 0 0
   Add 3 zeros.

4. 79 × 3
   __ 0 × _ = ____

5. 346 × 5
   ___ × _ = ____

6. 8,572 × 7
   ____ × _ = _____

7. 439 × 2
   ___ × _ = ____

8. 6,470 × 8
   ____ × _ = _____

Rounding to Multiply

For a quick, reasonable estimate:

• Round both factors to the lead digit.

• Multiply the front-end digits.

• Write as many zeros in the answer as there are in both factors.

Example	Step 1	Step 2
Estimate 38 × 56.	Round 38 to 40, and round 56 to 60.	Multiply 4 × 6, and add 2 zeros.
	40 × 60	40 × 60 = 2,400
		Add 2 zeros.

▶ Estimate by rounding both factors to the lead digit, then multiplying.

Round up ↓   ↓ Round down

1. 88 × 72 rounds to

_0_ × _0_ = _ , _0 0_

Add 2 zeros.

2. 52 × 29 rounds to

__ × __ = ____

3. 78 × 61 rounds to

__ × __ = ____

4. 27 × 13 rounds to

__ × __ = ____

5. 39 × 35 rounds to

_0_ × _0_ = _ , _0 0_

Add 2 zeros.

6. 24 × 41 rounds to

__ × __ = ____

7. 56 × 82 rounds to

__ × __ = ____

8. 92 × 23 rounds to

__ × __ = ____

Rounding to One-Digit Accuracy

When rounding larger numbers:

• Round both factors to the lead digit.

• Multiply the front-end digits.

• Write as many zeros in the answer as there are in both factors.

Example A

736 × 58 rounds to

700 × **6**0 = **42,**000

Multiply 7 × 6, and add 3 zeros.

Example B

596 × 345 rounds to

600 × **3**00 = **180,**000

Multiply 6 × 3, and add 4 zeros.

▶ Estimate by rounding both factors to the lead digit, then multiplying.

1. 419 × 19 rounds to

400 × 20 = _____

5. 432 × 370 rounds to

___ × ___ = _____

2. 573 × 63 rounds to

___ × ___ = _____

6. 114 × 715 rounds to

___ × ___ = _____

3. 940 × 22 rounds to

___ × ___ = _____

7. 501 × 638 rounds to

___ × ___ = _____

4. 354 × 51 rounds to

___ × ___ = _____

8. 271 × 513 rounds to

___ × ___ = _____

Is the Estimate High or Low?

When multiplying, it is sometimes easy to decide whether an estimate is high or low.

Estimate is high

If both factors are rounded up, the estimate will be high (overestimate).

Example 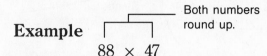 Both numbers round up.

88×47

Rounds to: 90×50
Estimate: 4,500 — high

Estimate is low

If both factors are rounded down, the estimate will be low (underestimate).

Example Both numbers round down.

24×63

Rounds to: 20×60
Estimate: 1,200 — low

Difficult to tell

If one factor is rounded up and the other down, it is difficult to tell whether the estimate is high or low.

Example
round down   round up

72×29

Rounds to: 70×30
Estimate: 2,100 — difficult to tell

▶ Round each factor, and estimate the product. Then circle whether the estimate is high, low, or difficult to tell.

1. 87×67

   Estimate: _____

   a) high
   b) low
   c) difficult to tell

2. 23×91

   Estimate: _____

   a) high
   b) low
   c) difficult to tell

3. 42×58

   Estimate: _____

   a) high
   b) low
   c) difficult to tell

4. 326×48

   Estimate: _____

   a) high
   b) low
   c) difficult to tell

5. 63×409

   Estimate: _____

   a) high
   b) low
   c) difficult to tell

6. 682×790

   Estimate: _____

   a) high
   b) low
   c) difficult to tell

Estimating with a Map

Knowing the distance between 2 cities can help you estimate other distances.

▶ The distance from New York to Chicago is about 800 miles. Use this fact and the map above to help you estimate the following distances.

1. Chicago to Denver _____
2. Chicago to Miami _____
3. New York to Houston _____

4. Houston to Seattle _____
5. Denver to Los Angeles _____
6. Seattle to New York _____

_____

▶ It takes about 2 hours to fly from Atlanta to Miami. Use this fact and the map above to help you estimate the following flight times.

7. Atlanta to Fargo _____
8. New York to Denver _____
9. Seattle to Los Angeles _____

10. Los Angeles to New York _____
11. Seattle to Miami _____
12. Houston to Fargo _____

Clustering

In some problems, a group of numbers clusters around a common value. When this happens, it is easy to arrive at a fast, reasonable estimate by using multiplication.

Weekly Wages	
Week	**Wages**
1	$296
2	305
3	284
4	320

Example

Estimate the total wages for 4 weeks.

The weekly wages all cluster around $300. Therefore, a reasonable estimate would be

$$4 \times \$300 = \$1,200$$

▶ For each problem, find the common value that the numbers cluster around. Then find your estimated answer.

1. Estimate the total miles driven in 4 days.

Miles Driven	
Day	**Miles**
1	485
2	520
3	492
4	505

a) What common value (average) does each number cluster around? _____500_____

b) Estimate:

_____

2. Estimate the total sales for 5 days.

Magazine Sales	
Day	**Sales**
1	$778
2	809
3	760
4	820
5	815

a) What common value (average) does each number cluster around? _____

b) Estimate:

_____

3.
$$\begin{array}{r} 4,876 \\ 5,050 \\ 4,969 \\ +5,025 \end{array}$$

a) What common value (average) does each number cluster around? _____

b) Estimate:

_____

4.
$$\begin{array}{r} 9,078 \\ 8,885 \\ 9,278 \\ 8,915 \\ +9,102 \end{array}$$

a) What common value (average) does each number cluster around? _____

b) Estimate:

_____

5.
$$\begin{array}{r} 10,189 \\ 9,961 \\ 10,075 \\ 9,850 \\ 10,283 \\ + 9,789 \end{array}$$

a) What common value (average) does each number cluster around? _____

b) Estimate:

_____

Does the Answer Make Sense?

When you are dividing, estimation helps you find a value that is close to the exact answer.

▶ Circle the answer that makes the most sense.

$295

1. Ann bought a bike for $295. She paid for it in 5 equal monthly payments. About how much did she pay each month?

   a) $6   **b)** $60   **c)** $600

2. 3 buses hold 147 students. About how many students will 1 bus hold?

   a) 15   **b)** 30   **c)** 50

3. The dinner bill of $78 was split 4 ways. About how much did each person pay?

   a) $2   **b)** $20   **c)** $200

4. Jill took out a car loan of $9,945 for 48 months. If she makes equal monthly payments, about how much will she pay each month?

   a) $50   **b)** $200   **c)** $500

5. Mr. Hansel traveled 415 miles on 20 gallons of gasoline. About how many miles did he travel per gallon?

   a) 20   **b)** 100   **c)** 200

Estimating with Division

Sometimes all we need is a rough estimate to know whether an answer is reasonable or not. When we divide, the location of the first digit will tell you whether the answer (**quotient**) is in the ones, tens, hundreds, thousands, and so on.

$9,856
48 months to pay

About how much is each monthly car payment?

Example

$$48\overline{)9,856}$$

Step 1

Think: Does 48 go into 9? No.

Does 48 go into 98? Yes.

Step 2

Place Xs over the 8, 5, and 6 to show how many digits are in the answer.

$$48\overline{)9,\overset{\text{X X X}}{8}56}$$

There will be three digits in the quotient. Therefore, the answer will be in the hundreds.

▶ Draw Xs to show the number of digits in each answer. Tell whether the answer will be in the ones, tens, hundreds, or thousands.

$$12\overline{)3,\overset{\text{X X X}}{5}67}$$

1. a) __ digits in the quotient
   b) The answer will be in the

   _____ .

$$8\overline{)496}$$

2. a) __ digits in the quotient
   b) The answer will be in the

   _____ .

$$28\overline{)1,932}$$

3. a) __ digits in the quotient
   b) The answer will be in the

   _____ .

$$3\overline{)6,318}$$

4. a) __ digits in the quotient
   b) The answer will be in the

   _____ .

Estimate the First Digit

Estimating the first digit in the quotient will help you make a reasonable estimate.

Example A

Change X to zero.

$$7)\overline{513}$$ with $7X$ above

Think:

7 will not go into 5.
7 will go into 51 (7 times).
Estimate: 70

Example B

Change Xs to zeros.

$$4)\overline{3,495}$$ with $8XX$ above

Think:

4 will not go into 3.
4 will go into 34 (8 times).
Estimate: 800

▶ Find the first digit in the quotient. Then estimate the quotient using zeros.

Change Xs to zeros.

1. $6)\overline{4,509}$ with $7XX$ above → Estimate: _____

2. $9)\overline{853}$ with $\square X$ above → Estimate: _____

3. $5)\overline{3,358}$ → Estimate: _____

4. $3)\overline{14,485}$ → Estimate: _____

5. $6)\overline{256}$ → Estimate: _____

6. $7)\overline{4,571}$ → Estimate: _____

7. $2)\overline{4,418}$ → Estimate: _____

8. $4)\overline{860}$ → Estimate: _____

9. $9)\overline{109}$ → Estimate: _____

10. $6)\overline{504}$ → Estimate: _____

Estimates Hold the Key

Rounding the **divisor** (the number you divide by) to the lead digit will help you estimate the quotient.

Example A

Round the divisor, 32, to 30.

Change Xs to zeros.

$$6XX$$
$$32\overline{)18,415}$$

round
32 } 30

Think:

30 will not go into 1.
30 will not go into 18.
30 will go into 184 (6 times).
Estimate: 600

Example B

Round the divisor, 579, to 600.

Change X to zero.

$$4X$$
$$579\overline{)26,571}$$

round
579 } 600

Think:

600 will not go into 2.
600 will not go into 26.
600 will not go into 265.
600 will go into 2,657 (4 times).
Estimate: 40

▶ Round the divisor to the lead digit. Then estimate the quotient.

Change Xs to zeros.

1.
$$\boxed{6}XX$$
$$53\overline{)32,468}$$

round
53 } 50   Estimate: _____

5.
$$44\overline{)86,987}$$

round
44 } _____   Estimate: _____

2.
$$\square X$$
$$487\overline{)15,874}$$

round
487 } 500   Estimate: _____

6.
$$86\overline{)648}$$

round
86 } _____   Estimate: _____

3.
$$67\overline{)16,130}$$

round
67 } _____   Estimate: _____

7.
$$607\overline{)38,241}$$

round
607 } _____   Estimate: _____

4.
$$759\overline{)58,531}$$

round
759 } _____   Estimate: _____

8.
$$12\overline{)71,807}$$

round
12 } _____   Estimate: _____

Compatible Numbers to Divide

Sometimes it is easier to estimate by using compatible numbers. **Compatible numbers** are numbers that divide exactly. When using compatible numbers to divide, think of basic division facts. Some basic division facts are:

$$5\overline{)40}^{\,8},\ 6\overline{)18}^{\,3},\ 9\overline{)36}^{\,4},\ \text{and}\ 4\overline{)28}^{\,7}$$

Example

5 and 36 are not compatible. 5 and 35 are compatible.

Replace Xs with zeros.

$$5\overline{)3,6\boxed{2}7} \qquad\qquad 5\overline{)3,5\boxed{2}7}\quad \dfrac{7\,\text{XX}}{}\to \dfrac{700}{\text{estimate}}$$

▶ Find the estimates by using compatible numbers.

1. 4 and 19 are not compatible. → **a)** 4 and _____ are compatible.

$$4\overline{)1,9\boxed{6}2}$$

b) $4\overline{)\_,\_62}\ \dfrac{\_\,\text{XX}}{}\to \dfrac{\_\_\_}{\text{estimate}}$

2. 7 and 40 are not compatible. → **a)** 7 and _____ are compatible.

$$7\overline{)4,0\boxed{6}5}$$

b) $7\overline{)\_,\_65}\ \to \dfrac{\_\_\_}{\text{estimate}}$

3. 6 and 28 are not compatible. → **a)** 6 and _____ are compatible.

$$6\overline{)2,8\boxed{0}2}$$

b) $6\overline{)\_,\_02}\ \to \dfrac{\_\_\_}{\text{estimate}}$

4. 9 and 51 are not compatible. → **a)** 9 and _____ are compatible.

$$9\overline{)5,1\boxed{3}8}$$

b) $9\overline{)\_,\_38}\ \to \dfrac{\_\_\_}{\text{estimate}}$

Apply Your Skills

▶ Use your estimation skills to make a reasonable estimate for each of the following questions.

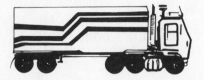

1. If 3 trucks can haul 2,694 boxes, about how many boxes can 1 truck haul? _____

2. Mike has a car loan of $16,284 for 48 months. If he makes equal monthly payments, about how much will he pay each month? _____

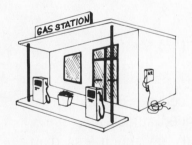

3. If Mr. Pavi drove 314 miles on 12 gallons of gasoline, about how many miles per gallon did he average? _____

4. A computer costs $2,595. If Stefania wants to save $290 each month, about how many months must she save to buy the computer? _____

5. Ann's 5-day cruise cost $1,684. About how much per day did it cost? _____

Review of Whole Number Estimation

▶ Answer each question by estimating. Be flexible. There are many ways to estimate. Before you start a problem, take a minute to think about the numbers involved. Use a method that is quick and easy to do in your head.

1. Estimate Matt's household expenses for January through April. _____

Household Expenses	
Month	**Expenses**
January	$107
February	98
March	110
April	92

2. Pete wants to save $165 each month. About how much will he save in 4 months? _____

3. The total cost of Kathy's stereo is $518 to be paid in 5 equal monthly payments. About how much is each payment? _____

4. Magazine sales were $3,486 on Monday and $1,560 on Tuesday. Monday's sales were about how much more than Tuesday's? _____

5. Donna paid $295 each month for 6 months on her car loan. About how much did she pay altogether? _____

6. Greg traveled 438 miles on 21 gallons of gasoline. About how many miles did he travel on each gallon of gasoline? _____

7. Home A was priced at $87,645. Home B was priced at $68,108. Home B is about how much less than Home A? _____

8. About what is the total cost of the 3 items? _____

Item	Cost
Headset	$ 38
Jacket	145
Stereo System	295

Estimating with Decimals

Listed below are exact numbers paired with their approximate values (values close to the original numbers). Notice the key words that show the values aren't exact. Approximate values are often used because they're easier and quicker to work with.

	Exact Numbers	Approximate Values
Portable TV	$399.99	**Nearly** $400
Kareem's height	2.1 meters	**About** 2 meters
Men's jeans	$18.98	**Almost** $20.00
Lawn tractor	$975.99	**Close to** $1,000
Roast beef	3.2 pounds	A **little over** 3 pounds
Photo album	$.96	A **little less** than $1.00

▶ Estimate the answers using approximate amounts of money.

$1.98

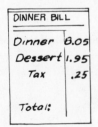

$2.30

1. Will $5 be enough to buy both items? _____

2. Will $10 be enough to pay for the dinner bill? _____

$6.12

$2.98

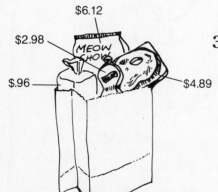

$.96

$4.89

3. Will $20 be enough to pay for the groceries? _____

Front-End Estimation

As you learned in the section on whole numbers, front-end estimation is a method in which you add together the digits with the largest place values (front-end digits).

Example A	Example B	Example C
$4.27 8.45 +5.19 **17** ones = $17 estimate	$27.18 14.50 32.29 +89.37 **14** tens = $140 estimate	$734.15 225.29 867.30 +414.45 **21** hundreds = $2,100 estimate

▶ Add the front-end digits to get a reasonable estimate.

1. $3.15
   +9.63
   __12__ ones = $ _____
   estimate

2. $3.15
   9.95
   +5.25
   ___ ones = $ _____
   estimate

3. $3.08
   7.37
   4.65
   +2.62
   ___ ones = $ _____
   estimate

4. $16.63
   47.22
   +65.87
   ___ tens = $ _____
   estimate

5. $38.91
   Think zero. → 7.36
   71.39
   +80.52
   ___ tens = $ _____
   estimate

6. $34.28
   98.31
   76.60
   22.59
   +10.74
   ___ tens = $ _____
   estimate

7. $372.92
   Think zero. → 30.67
   +425.38
   ___ hundreds = $ _____
   estimate

8. $588.94
   310.49
   957.85
   +703.66
   ___ hundreds = $ _____
   estimate

9. $329.40
   455.53
   71.89
   283.46
   +846.07
   ___ hundreds = $ _____
   estimate

Rounding to Lead Digits

When estimating in real life, we often round money amounts to the greatest place value (lead digit). To round a number to the lead digit, look at the next greatest place value to the right.

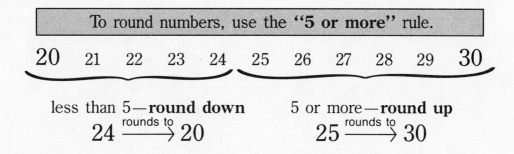

To round numbers, use the **"5 or more"** rule.

20 21 22 23 24 25 26 27 28 29 30

less than 5—**round down**
$24 \xrightarrow{\text{rounds to}} 20$

5 or more—**round up**
$25 \xrightarrow{\text{rounds to}} 30$

If the digit to the right of the lead digit is less than 5 (4, 3, 2, 1, 0), change all the numbers to 0 except the lead digit.

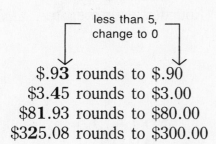

less than 5, change to 0

$.9\underline{3}$ rounds to $.90
$3.\underline{4}5$ rounds to $3.00
$8\underline{1}.93$ rounds to $80.00
$3\underline{2}5.08$ rounds to $300.00

If the digit to the right of the lead digit is 5 or more (5, 6, 7, 8, 9), add 1 to the lead digit. Then change all the other digits to 0.

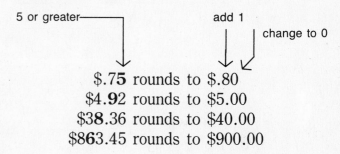

5 or greater   add 1   change to 0

$.\underline{7}5$ rounds to $.80
$4.\underline{9}2$ rounds to $5.00
$3\underline{8}.36$ rounds to $40.00
$8\underline{6}3.45$ rounds to $900.00

▶ Based on the examples shown above, circle the correct rounded number from the 2 choices.

1. $7.98 rounds to $7.00 or ($8.00)

2. $.54 rounds to $.50 or $.60

3. $549.72 rounds to $500.00 or $600.00

4. $25.98 rounds to $20.00 or $30.00

5. $.98 rounds to $.90 or $1.00

6. $4.69 rounds to $4.00 or $5.00

7. $103.25 rounds to $100.00 or $200.00

▶ Round each money amount to the lead digit (the first digit on the far left).

8. $29.45 rounds to _____

9. $.65 rounds to _____

10. $529.62 rounds to _____

11. $1.19 rounds to _____

12. $46.80 rounds to _____

13. $22.68 rounds to _____

14. $641.75 rounds to _____

Add by Rounding to Lead Digits

A quick way to estimate money amounts is to round the lead digits.

Example A

$18.95 rounds to $20.00
7.68 rounds to   8.00
+ 63.25 rounds to   60.00
Rounded estimate: $88.00

Example B

$244.27 rounds to $200.00
395.09 rounds to   400.00
+ 49.98 rounds to   50.00
Rounded estimate: $650.00

▶ Round each dollar amount to the lead digit. Then estimate the answers.

1. $27.98 rounds to $ _30.00_
    73.19 rounds to _____
    + 3.82 rounds to + _____
    Rounded estimate: $ _____

4. $283.92
    554.17
    + 88.60
    Rounded estimate: $ _____

2. $ 3.75 rounds to $ _____
    38.28 rounds to _____
    + 4.77 rounds to + _____
    Rounded estimate: $ _____

5. $614.93
    989.05
    + 73.16
    Rounded estimate: $ _____

3. $ 5.47
    9.58
    + 2.82
    Rounded estimate: $ _____

6. $ 48.71
    29.84
    + 9.39
    Rounded estimate: $ _____

Estimating Dollar Amounts

Another method for estimating money amounts is to look for amounts that add up to about $1.00.

▶ Find the estimate for each problem by following these steps:
  • First, connect the pair of prices that "go together" to make about $1.00.
  • Then add the remaining price for your estimate.

1. $.43
   About $1.00
   $.62 $.15

Estimate: $1.15

2. $.48
   $.75 $.49

Estimate: _____

3. $.65
   $.08 $.88

Estimate: _____

4. $.39
   $.57 $.05

Estimate: _____

5. $.55
   $.19 $.79

Estimate: _____

6. $.45
   $.58 $.20

Estimate: _____

7. $.12
   $.50 $.90

Estimate: _____

8. $.29
   $.51 $.48

Estimate: _____

Adjusting Front-End Estimates

We can make front-end estimates of money amounts more accurate by adjusting them.

▶ Find the adjusted estimate for each problem by following these steps:

• First, add the front-end digits.

• Next, group the cents into dollars whenever possible.

• Then add the amounts found in the first 2 steps to find the adjusted estimate.

1. $1.29 ← about $1.00
    3.67
    2.49 ← about $1.00
   +5.55

   a) $11 and about $ __ more

   b) Adjusted estimate: $ _____

2. $ .98 ← about $1.00
    7.58
    2.29 ← about $1.00
   +3.09

   a) $ _____ and about $ __ more

   b) Adjusted estimate: $ _____

3. $3.73
     .47 ⎫ about $1.00
    6.24 ⎬ about $1.00
   + .49 ⎭

   a) $ _____ and about $ __ more

   b) Adjusted estimate: $ _____

4. $7.08
     .88
     .34
   + .62

   a) $ _____ and about $ __ more

   b) Adjusted estimate: $ _____

5. $2.19
    4.27
    7.09
   +1.42

   a) $ _____ and about $ __ more

   b) Adjusted estimate: $ _____

6. $8.99
    7.48
    5.39
   +2.08

   a) $ _____ and about $ __ more

   b) Adjusted estimate: $ _____

7. $5.63
     .46
    2.34
   + .55

   a) $ _____ and about $ __ more

   b) Adjusted estimate: $ _____

8. $9.99
    8.17
     .80
   +3.98

   a) $ _____ and about $ __ more

   b) Adjusted estimate: $ _____

Estimate Only Dollar Amounts

For large amounts of money (greater than $10), it is not important to estimate the cents. Practice your estimation skills using only the dollar amounts.

Example

$105.75
$228.95
+$476.92

Estimate using only dollar amounts.

$105
$228
+$476

While all 3 types of estimation methods are useful, methods 2 and 3, shown below, will help you find an estimate closer to the exact answer.

Method 1	Method 2	Method 3
Add the front-end digits.	Adjust the front-end estimate.	Round to the lead digits.

Method 1:

$1 05
2 28
+4 76
7 hundred = $700

Method 2:

$1|05|
2|28| ← about $100
+4|76|
$700 and about $100 more
Adjusted estimate: $800

Method 3:

$1 05 rounds to $100
2 28 rounds to 200
+4 76 rounds to 500
Estimate: $800

▶ Use the front-end, adjusting, and rounding methods to estimate. Use only dollar amounts.

1. $731.85
343.26
+524.98

  a) Front-end: $1,500
  b) Adjusted: _____
  c) Rounded: _____

3. $798.06
473.42
+232.13

  a) Front-end: _____
  b) Adjusted: _____
  c) Rounded: _____

5. $76.18
16.07
+59.72

  a) Front-end: _____
  b) Adjusted: _____
  c) Rounded: _____

2. $89.52
54.89
+18.75

  a) Front-end: _____
  b) Adjusted: _____
  c) Rounded: _____

4. $833.29
+362.83

  a) Front-end: _____
  b) Adjusted: _____
  c) Rounded: _____

6. $615.07
721.95
+360.33

  a) Front-end: _____
  b) Adjusted: _____
  c) Rounded: _____

Planning a Picnic

Murch's Picnics to Go

4 Ham Sandwiches $18.75	2 lbs.* Baked Beans $2.05
4 Turkey Sandwiches 12.95	1 pack Cola 2.89
2 bags Potato Chips 4.15	Plates & Napkins 2.15
2 lbs.* Cole Slaw 1.95	Cookies 1.09

*lbs. = pounds

▶ You are planning a picnic and want to estimate some basic takeout prices. To answer the questions, use the money amounts for the items listed below.

1. Will $10 be enough to buy the following items?

    2 bags potato chips

    2 lbs. cole slaw

    1 pack cola

a) yes   **b)** no

2. About how much will it cost to buy the following items?

    4 ham sandwiches

    2 bags potato chips

    1 pack cola

Estimate: _____

3. Will $20 be enough to buy the following items?

    4 turkey sandwiches

    2 bags potato chips

    2 lbs. cole slaw

a) yes   **b)** no

4. About how much will you pay for the 3 least expensive items? _____

5. About how much will you pay for the 2 most expensive items? _____

6. Will $5 be enough to buy 2 lbs. of baked beans and 1 pack of cola?

a) yes   **b)** no

7. Will $50 be enough to buy all the items?

a) yes   **b)** no

Is It Sensible?

▶ Circle the estimate that makes the most sense.

Regular price—$84.62
Sale price—$53.95

1. About how much is saved by buying the tennis racket on sale?

   a) $20   **b)** $30   **c)** $40

$3.62 $1.79

2. About what is the difference in price between the two boxes?

   a) $5   **b)** $2   **c)** $10

3. About how much change will you get back from $100?

   a) $5   **b)** $10   **c)** $25

4. About how much change will you get back from $20?

   a) $15   **b)** $20   **c)** $25

Total $4.36

Rounding Decimals to Subtract

Rounding decimal numbers to their lead digits makes it easy to subtract in your head.

Example A

5 or greater, round up

$4.87	rounds to	$5
− 1.95	rounds to	−2

Rounded estimate: $3

Example B

4 or less, round down

$74.37	rounds to	$70
− 10.85	rounds to	−10

Rounded estimate: $60

▶ Round each decimal number to the lead digit, then estimate the difference.

1. $7.23   rounds to __$7__
    − 3.58   rounds to __−4__
    Rounded estimate: _____

5. $61.94
    − 18.32
    Rounded estimate: _____

2. $92.46   rounds to _____
    − 39.79   rounds to _____
    Rounded estimate: _____

6. $673.83
    − 316.09
    Rounded estimate: _____

3. $72.45   rounds to _____
    − 28.98   rounds to _____
    Rounded estimate: _____

7. $9.15
    − 4.83
    Rounded estimate: _____

4. $867.90   rounds to _____
    − 273.50   rounds to _____
    Rounded estimate: _____

8. $53.83
    − 12.19
    Rounded estimate: _____

Front-End Estimation and Subtraction

To estimate using the front-end method:

• Subtract the front-end digits.

• Compare the remaining digits for a closer estimate.

Example A

Step 1

Subtract the front-end digits (8, 2) to find the front-end estimate.

$8.84
−2.15
6 ones = $6

Step 2

Compare the remaining numbers, 84 and 15.

$8.|84|
−2.|15|

← The top number is larger than the bottom number, so the answer is **greater than** $6.

Greater than $6

Example B

Step 1

Subtract the front-end digits (9, 5) to find the front-end estimate.

$9.24
−5.82
4 ones = $4

Step 2

Compare the remaining numbers, 24 and 82.

$9.|24|
−5.|82|

← The top number is smaller than the bottom number, so the answer is **less than** $4.

Less than $4

▶ Follow the steps shown above to find a close front-end estimate. Circle the answer.

1. $7.46
   −1.98

   a) greater than $6

   b) less than $6

2. $5.43
   −2.08

   a) greater than $3

   b) less than $3

3. $6.29
   −2.36

   a) greater than $4

   b) less than $4

4. $3.47
   −2.91

   a) greater than $1

   b) less than $1

5. $8.87
   −6.15

   a) greater than $2

   b) less than $2

6. $8.43
   −1.15

   a) greater than $7

   b) less than $7

Adjusting Front-End Estimation

When subtracting decimals, sometimes you will want to get a closer estimate.

Example A

Step 1

Find the front-end estimate.

$$\begin{array}{r} \$83.68 \\ -39.25 \\ \hline \end{array}$$

5 tens = $50

Step 2

To find the adjusted estimate, **you must regroup because 9 is greater than 3.** (Use only the dollar amounts.)

$$\begin{array}{r} \overset{7\ 13}{\$8\cancel{3}.68} \\ -39.25 \\ \hline \end{array}$$

44 dollars

Example B

Step 1

Find the front-end estimate.

$$\begin{array}{r} \$77.24 \\ -34.63 \\ \hline \end{array}$$

4 tens = $40

Step 2

To find the adjusted estimate, no regrouping is necessary because 4 is less than 7. (Use only the dollar amounts.)

$$\begin{array}{r} \$77.24 \\ -34.63 \\ \hline \end{array}$$

43 dollars

▶ Find the front-end and adjusted estimates. Use only the dollar amounts.

1. $$\begin{array}{r} \$54.65 \\ -38.83 \\ \hline \end{array}$$

    a) Front-end estimate: ___$20___

    b) Adjusted estimate: _____

2. $$\begin{array}{r} \$76.51 \\ -31.75 \\ \hline \end{array}$$

    a) Front-end estimate: _____

    b) Adjusted estimate: _____

3. $$\begin{array}{r} \$382.28 \\ -142.73 \\ \hline \end{array}$$

    a) Front-end estimate: _____

    b) Adjusted estimate: _____

4. $$\begin{array}{r} \$62.48 \\ -29.02 \\ \hline \end{array}$$

    a) Front-end estimate: _____

    b) Adjusted estimate: _____

5. $$\begin{array}{r} \$728.89 \\ -455.62 \\ \hline \end{array}$$

    a) Front-end estimate: _____

    b) Adjusted estimate: _____

6. $$\begin{array}{r} \$94.48 \\ -51.03 \\ \hline \end{array}$$

    a) Front-end estimate: _____

    b) Adjusted estimate: _____

Spring Specials

Estimates are useful when you want to get a rough idea of costs and savings when shopping.

Regular Price—$379.95
Sale Price—$296.95

Regular Price—$67.89
Sale Price—$53.99

Regular Price—$14.98
Sale Price—$ 9.79

Regular Price—$89.99
Sale Price—$68.75

▶ Using the prices of the items pictured, find estimates for the questions below. All reasonable answers are acceptable.

1. About how much do you save by buying each item on sale?

   a) lawn mower: _____

   b) running jacket: _____

   c) running shorts: _____

   d) outdoor grill: _____

2. If Perry has a balance of $936.29 in his savings account and buys the lawn mower at the sale price, about how much will he have left in his savings account? _____

3. About how much change will Sally get back from $80 if she buys the running jacket and shorts at the sale price? _____

4. The price of the lawn mower on sale is about how much more than the price of the outdoor grill on sale? _____

Estimates That Make Sense

It is important to make a quick estimate of the amount of change you should get after a purchase.

▶ Compare the amount of the purchase and the amount given to the cashier. Circle the most reasonable estimate of the amount of change you should receive.

Paid $5.00
for Each Purchase

	Amount of Purchase	Amount of Change		
1.	$2.43	a) $1.25	b) $2.60	c) $.95
2.	$1.19	a) $1.25	b) $2.75	c) $3.80
3.	$0.78	a) $4.25	b) $1.70	c) $2.30

Paid $20.00
for Each Purchase

	Amount of Purchase	Amount of Change		
4.	$4.87	a) $10.00	b) $15.00	c) $5.00
5.	$9.05	a) $11.00	b) $5.00	c) $15.00
6.	$18.97	a) $10.00	b) $5.00	c) $1.00

Paid $100.00
for Each Purchase

	Amount of Purchase	Amount of Change		
7.	$13.99	a) $85.00	b) $45.00	c) $15.00
8.	$82.10	a) $48.00	b) $18.00	c) $8.00
9.	$48.55	a) $25.00	b) $50.00	c) $75.00

Shopping Trips

When shopping, you can use estimation to quickly check whether or not you have enough money to buy what you want.

▶ Use one of the estimation methods that you learned earlier to answer the following questions.

Item	Cost
Battery	$58.89
Gasoline	9.75
Oil	4.95

1. Will $70 be enough to buy all the items? _____

2. About how much change will you get back from $70 if you decide to buy only the battery? _____

Item	Cost
Rolls	$2.89
Coffee	.95
Bread	3.28

3. Will $5 be enough to buy all the items? _____

4. About how much change will you get back from $5 if you buy just the rolls and coffee? _____

Item	Cost
Hammer	$ 9.89
Nails	5.09
Paint	18.98
Tape measure	5.95

5. Will $50 be enough to buy all the items? _____

6. About how much change will you get back from $50 if you buy the hammer, nails, and paint? _____

Be Reasonable

▶ Reasonable answers require good number sense. Circle the answer that makes the most sense.

$2.98

1. About how much will 4 paintbrushes cost?

    a) $6   **b)** $12   **c)** $18

Monthly Bill
Cable TV
$19.48

2. About how much will 6 months of cable TV cost?

    a) $180   **b)** $60   **c)** $120

$9.88

Garden hose

3. About how much will 3 garden hoses cost?

    a) $30   **b)** $20   **c)** $10

4. About how much will 5 pounds of peanuts cost?

    a) $3   **b)** $5   **c)** $7

Round the Dollar Amounts

For a quick estimate, round only the dollar amounts to the lead digits. It is not necessary to round a one-digit number.

Example	Step 1	Step 2
About how much will 4 shirts cost?	Round the dollar amount to the lead digit.	Multiply the number of items by the rounded cost.

 $6.89

$6.89 rounds to $7

↑ 5 or greater, round up

$$4 \times \$7 = \$28$$

4 shirts will cost about $28.

▶ Estimate the costs by rounding the dollar amounts to the lead digits before multiplying.

 $8.28

 $2.99

1. About how much will 3 pairs of shorts cost?

a) Round the dollars: _____
b) _____ × _____
c) 3 pairs of shorts will cost about _____ .

3. About how much will 6 pairs of socks cost?

a) Round the dollars: _____
b) _____ × _____
c) 6 pairs of socks will cost about _____ .

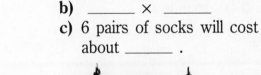 House Payment—$415.74

2. About how much will 5 tickets cost?

a) Round the dollars: _____
b) _____ × _____
c) 5 tickets will cost about _____ .

4. About how much will 3 house payments cost?

a) Round the dollars: _____
b) _____ × _____
c) 3 house payments will cost about _____ .

Multiplying Mixed Decimals

Rounding mixed decimals to lead digits can help a shopper estimate costs.

Example A

Roast beef costs $3.87 per pound. About how much will 2.3 pounds cost?

$3.87 rounds to $4
× 2.3 rounds to ×2
Estimate: $8

2.3 pounds will cost about $8.

Example B

1 pen costs $2.29. About how much will 5 pens cost?

$2.29 rounds to $2
×   5 ———→ ×5 ← It is not necessary to round a 1-digit factor.
Estimate: $10

5 pens will cost about $10.

▶ Estimate the answers by rounding the mixed decimals to their lead digits.

1. $1.93 rounds to $ _2_
    × 2.7 rounds to × _3_
    Estimate: _____

4. $5.09 rounds to $_____
    ×   8 ———→ ×_____
    Estimate: _____

2. $4.85 rounds to $_____
    × 3.4 rounds to ×_____
    Estimate: _____

5. $1.69 rounds to $_____
    ×   2 ———→ ×_____
    Estimate: _____

3. $8.79 rounds to $_____
    × 4.6 rounds to ×_____
    Estimate: _____

6. $9.38 rounds to $_____
    ×   7 ———→ ×_____
    Estimate: _____

Buying Fruits and Vegetables

Estimation is a useful skill when shopping.

Example

Bananas

$.29 per pound

About how much will 6 pounds cost?

6 × $.29 per pound

| rounds to |

6 × $.30 = $1.80

6 pounds of bananas will cost about $1.80.

▶ Estimate the cost by rounding the money amount to the lead digit and multiplying by the number of items.

1.   Tomatoes

$.47 per pound

About how much will 5 pounds of tomatoes cost?

5 × $.47

| rounds to |

5 × _____ = _____

5 pounds of tomatoes will cost about _____ .

---

2.   Onion Bunches

$.78 per bunch

About how much will 2 bunches of onions cost?

2 × $.78

| rounds to |

2 × _____ = _____

2 onion bunches will cost about _____ .

---

3.   Apples

$.55 per pound

About how much will 8 pounds of apples cost?

8 × $.55

| rounds to |

8 × _____ = _____

8 pounds of apples will cost about _____ .

Getting Closer

Sometimes you may want to get a closer estimate when multiplying with 1-digit factors.

Example

About how much will 4 cans of tennis balls cost?

$3.48

Step 1	Step 2	Step 3
Multiply the dollars by the number of items.	Round the cents, and multiply.	Add the two products.

Step 1
$4 \times \$3.48$
$4 \times \$3 = \$12$

Step 2
rounds to $.50
$4 \times \$3.\boxed{48}$
$4 \times \$.50 = \$2.00$

Step 3
$4 \times \$3 = \$12$
$4 \times \$.50 = \$2$ — add
Estimate: $14

▶ Follow the steps above to find close estimates.

 $4.29

 $3.67

1. About how much will 5 six-packs cost?

  a) Multiply the dollars $20

  b) Multiply the cents   _____

  c) Add   _____

3. About how much will 2 packages of batteries cost?

  a) Multiply the dollars _____

  b) Multiply the cents   _____

  c) Add   _____

 $1.83

 $8.64

2. About how much will 3 gallons of milk cost?

  a) Multiply the dollars _____

  b) Multiply the cents   _____

  c) Add   _____

4. About how much will 4 lawn chairs cost?

  a) Multiply the dollars _____

  b) Multiply the cents   _____

  c) Add   _____

Estimate the Costs

▶ Complete the tables using the information given and your estimating skills. All reasonable estimates are acceptable.

Item	Number of Pounds Bought	Price per Pound	Estimated Cost
Nut mix	2	**1. a)**	**b)**
Peanuts	1.8	**2. a)**	**b)**
Candy	4	**3. a)**	**b)**
Coffee	3.2	**4. a)**	**b)**

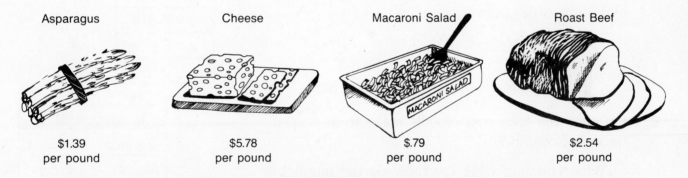

Asparagus — $1.39 per pound

Cheese — $5.78 per pound

Macaroni Salad — $.79 per pound

Roast Beef — $2.54 per pound

Item	Number of Pounds Bought	Price per Pound	Estimated Cost
Macaroni salad	4	**5. a)**	**b)**
Cheese	2.2	**6. a)**	**b)**
Asparagus	2	**7. a)**	**b)**
Roast Beef	3.4	**8. a)**	**b)**

Clustering

In some problems, a group of numbers cluster around a common value. When this happens, you can use the clustering method shown below.

Weekly Savings	
Week	Savings
1	$20.95
2	$18.25
3	$19.50
4	$22.30

Example

Estimate the total savings for 4 weeks.

The weekly savings all cluster around an average of $20.00

Therefore, a reasonable estimate is
4 × $20 = $80

▶ Use the information from the tables to estimate the answers.

1. Estimate the total miles traveled for the 5-day bicycle trip.

   a) What common value (average) do the numbers cluster around? _____

   b) Estimate the answer: _____

Bicycle Trip	
Day	Miles
1	29.2
2	30.9
3	27.5
4	32.4
5	31.6

2. Estimate the total cost of calls made from August through November.

   a) What common value (average) do the numbers cluster around? _____

   b) Estimate the answer: $_____

Telephone Bills	
Month	Amount
August	$ 8.95
September	11.15
October	10.28
November	9.83

3. $ 5,789.75
     6,218.83
     5,962.24
     5,814.55
   +6,156.88

   a) What common value (average) do the numbers cluster around? _____

   b) Estimate the answer: _____

Dividing to Estimate

Estimation is a skill we use every day. It requires making a reasonably accurate guess.

▶ Circle the most reasonable answer.

4 tires for $206.95

1. About what is the cost of 1 tire?

    a) $5   **b)** $50   **c)** $500

$21.08
Gas

2. If 17 gallons of gas cost $21.08, about what is the cost per gallon?

    a) $1.25   **b)** $5.25   **c)** $10.25

6 muffins
for $3.70

3. About how much does 1 muffin cost?

    a) $.06   **b)** $.60   **c)** $6.00

8 rose plants
for $56.64

4. About how much does 1 rose plant cost?

    a) $.07   **b)** $.70   **c)** $7.00

Estimating to Divide

Sometimes, when you divide a decimal by a 1-digit divisor, you can simply make a quick, rough estimate.

Example A

If 3 frozen pies cost $13.95, about how much does 1 cost?

Change Xs to zeros.

$$\frac{\$4.XX}{3)\$13\uparrow95}$$

3 will not go into 1.
3 will go into 13 (4 times).

Each pie costs about $4.00.

Example B

$385.70 is equally divided among 4 people. About how much money will each person receive?

Change Xs to zeros.

$$\frac{9X.XX}{4)\$385.70}$$

4 will not go into 3.
4 will go into 38 (9 times).

Each person will receive about $90.00.

▶ Divide mentally to find the first digit in the quotient. Then estimate your answer to the nearest dollar.

1.
$$\frac{4.XX}{7)\$29\uparrow75}$$ about $ _4.00_

2.
$$\frac{\_.XX}{2)\$10\uparrow79}$$ about $ _____

3.
$$4)\$38.95$$ about $ _____

4.
$$8)\$16.25$$ about $ _____

5.
$$9)\$279.29$$ about $ _____

6.
$$5)\$422.99$$ about $ _____

7.
$$3)\$312.16$$ about $ _____

8.
$$6)\$559.85$$ about $ _____

Two-Digit Divisors

Rounding the divisor to the lead digit will help you estimate the quotient.

Example

Special

24 exposures for $4.35

About how much will 1 exposure cost?
Round the divisor, 24, to 20.

Change X to zero.

$$.2X$$
$$24\overline{)\$4\uparrow35}$$

round 24 } 20

Think:
20 will not go into 4.
20 will go into 43 (2 times).
Estimate: $.20

▶ Round the divisor to the lead digit. Then estimate the quotient.

Change Xs to zeros.

1.
$$\square.XX$$
$$36\overline{)165\uparrow98}$$

round 36 } _40_   Estimate: _____

4.
$$88\overline{)659.82}$$

round 88 } _____   Estimate: _____

2.
$$.\_X$$
$$18\overline{)2\uparrow35}$$

round 18 } _____   Estimate: _____

5.
$$518\overline{)47,218.94}$$

round 518 } _____   Estimate: _____

3.
$$212\overline{)8,275.35}$$

round 212 } _____   Estimate: _____

6.
$$12\overline{)735.92}$$

round 12 } _____   Estimate: _____

Compatible Numbers to Divide

Sometimes you will get a closer estimate by using compatible numbers. When using compatible numbers to divide, think of basic division facts (numbers that divide exactly). Some basic division facts are: $5\overline{)30}$, $6\overline{)24}$, $7\overline{)28}$, and $9\overline{)63}$ giving $6, 4, 4, 7$

Example

7 and 33 are *not* compatible.   →   7 and 35 *are* compatible.

$7\overline{)\$33.53}$

$\dfrac{\$5.\text{XX}}{7\overline{)\$35\uparrow53}} \rightarrow \dfrac{\$5.00}{\text{estimate}}$

▶ In each problem below, find compatible numbers to work with. Then estimate the quotient.

1. 8 and 70 are not compatible.   →   **a)** 8 and _____ are compatible.

$8\overline{)\$70.36}$

b) $\dfrac{\$\_.\text{XX}}{8\overline{)\$\_\_\uparrow36}} \rightarrow \underline{}$ estimate

2. 5 and 23 are not compatible.   →   **a)** 5 and _____ are compatible.

$5\overline{)\$23.72}$

b) $\dfrac{\$\_.\_\_}{5\overline{)\$\_\_\uparrow72}} \rightarrow \underline{}$ estimate

3. 6 and 29 are not compatible.   →   **a)** 6 and _____ are compatible.

$6\overline{)\$29.15}$

b) $\dfrac{\$\_.\_\_}{6\overline{)\$\_\_\uparrow15}} \rightarrow \underline{}$ estimate

4. 7 and 27 are not compatible.   →   **a)** 7 and _____ are compatible.

$7\overline{)\$27.83}$

b) $\dfrac{\$\_.\_\_}{7\overline{)\$\_\_\uparrow83}} \rightarrow \underline{}$ estimate

Real-World Applications

▶ Estimate your answers. All reasonable answers are acceptable

Microwave

$238
12 months to pay

1. Estimate the payment for 1 month.

Estimate: _____

Television

$549.87
6 months to pay

2. Estimate the payment for 1 month.

Estimate: _____

Tires

4 for $435.95

3. About how much would 1 tire cost?

Estimate: _____

Dinner Bill

$26.95
split 3 ways

4. About how much would one-third of the dinner cost?

Estimate: _____

Candy

3 pounds for $8.75

5. About how much would 1 pound cost?

Estimate: _____

Sandwiches

5 for $18.95

6. About how much would 1 sandwich cost?

Estimate: _____

Estimate for Best Buy

To estimate the best value for money, compare the unit prices. The **unit price** is the cost for 1, or each, unit.

To find the unit price, divide the cost by the number of units of measure.

Example $2.49

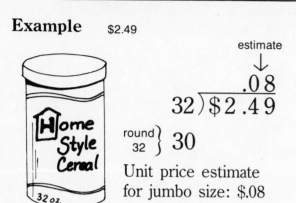

$$\begin{array}{r} \overset{\text{estimate}}{\downarrow} \\ .08 \\ 32\overline{)\$2.49} \end{array}$$

round
32 } 30

Unit price estimate
for jumbo size: $.08

$1.39

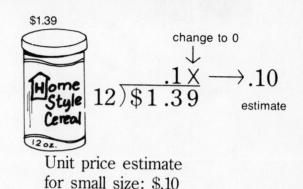

$$\begin{array}{r} \overset{\text{change to 0}}{\downarrow} \\ .1\text{X} \longrightarrow .10 \\ 12\overline{)\$1.39} \end{array}$$

estimate

Unit price estimate
for small size: $.10

Because $.08 per ounce is less than $.10 per ounce, the jumbo size at $2.49 has the lower unit price.

▶ To estimate unit prices, you may need to round the divisor as shown in the example above.

1. Which item has the lower unit price? _____

$12.95

Unit price estimate: _____

$4.85

Unit price estimate: _____

2. Does the 48-ounce or the 28-ounce box have the lower unit price? _____

$3.59

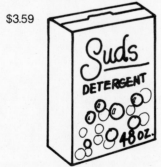

Unit price estimate: _____

$2.45

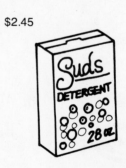

Unit price estimate: _____

Practice Your Skills

▶ Read each problem and decide which operation to use (addition, subtraction, multiplication, or division). Then estimate the answer.

1. The regular price of an item is $45.32. The sale price is $36.50. About how much could be saved by buying the item on sale? _____

2. 4 batteries cost $3.45. About how much does each battery cost? _____

3. A dinner bill of $37.45 was split 4 ways. About how much did each person pay? _____

4. If your car can travel 18.75 miles on 1 gallon of gasoline, about how many miles could it travel on 9 gallons? _____

▶ Use estimates, **not** exact calculations, to compare the numbers below. Place the symbols < (less than) or > (greater than) in the ⬤ to make each statement true.

5. $4.19 + $15.63 ⬤ $20.00

6. 7.14 × 5.2 ⬤ 35

7. $853.95 − $206.48 ⬤ $600.00

8. $29.52 ÷ 5 ⬤ $6.00

9. $251.79 + $48.95 ⬤ $270.00

10. $47.83 − $20.84 ⬤ $30.00

Ballpark Estimates

Many problems in daily life require only estimates.

▶ Take a minute to think about the numbers involved in each problem, then make an estimate. All reasonable answers are acceptable.

$15,898

$8,120

1. About what is the difference in price? _____

$1.54

2. About how much will 3 hamburgers cost? _____

$76,895

$98,985

3. About what is the difference in price? _____

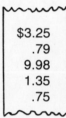

$3.25
.79
9.98
1.35
.75

4. About how much was spent at the grocery store? _____

5. Philip and a friend went on a 4-day bicycle ride. They rode the following distances:

1st day	21.6 miles
2nd day	20.9 miles
3rd day	19.4 miles
4th day	18.5 miles

About how many miles did they ride on their 4-day trip? _____

Orange Juice	Cereal	Bananas
$1.79	$2.50	$1.10

6. Can you buy the 3 pictured items for $5.00? _____

Reasonable Guess

When shopping we often make many guesses (estimates).

Examples:

A. Do I have enough money?
B. What is the better buy?
C. About how much will 3 pounds cost?

D. Can I buy 2 for $10?
E. Does the total cost seem about right?
F. About how much change will I get back?

Soda
$.59

Eggs
$1.19

Soup
3 cans for $2.78

Baked Ham
$3.98 per pound

Bread
$.92

▶ Based on the information given above, estimate to answer the questions below.

1. Will $5.00 be enough to buy 1 bottle of soda, 1 pound of baked ham, and 1 loaf of bread? _____

2. Will $10 be enough to buy 2 pounds of baked ham and a carton of eggs? _____

3. Can you buy 1 pound of ham and the rest of the pictured grocery items for less than $10? _____

4. About how much will 1 can of soup cost? _____

5. About how much will 8 bottles of soda cost? _____

6. If you buy 3 cans of soup and 2 pounds of baked ham, about how much change will you get back from $20.00? _____

Number Sense

Estimates are useful to check exact computations, even when you use a calculator. Calculators do not make mistakes, but sometimes the person using a calculator makes errors when entering numbers or decimal points.

▶ Make an estimate for each calculation. Then circle whether the answer shown on the calculator is reasonable.

1. 947 − 205

Estimate: _____

Is the calculator answer reasonable?

a) yes   **b)** no

4. 5,096 × 5

Estimate: _____

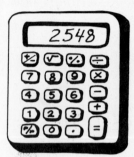

Is the calculator answer reasonable?

a) yes   **b)** no

2. $80.55 − 9

Estimate: _____

Is the calculator answer reasonable?

a) yes   **b)** no

5. $3.67 + $7.58 + $.85

Estimate: _____

Is the calculator answer reasonable?

a) yes   **b)** no

3. 4.15 × 6.8

Estimate: _____

Is the calculator answer reasonable?

a) yes   **b)** no

6. $82.45 − $7.15

Estimate: _____

Is the calculator answer reasonable?

a) yes   **b)** no

Review of Decimal Estimation

▶ Answer each question by estimating. Be flexible. There are many ways to estimate. Before you start a problem, take a minute to think about the numbers involved. Use a method that is quick and easy to do in your head.

1. Norma bought the following items. About how much did she spend altogether? _____

        Blouse: $24.95

        Pants: $38.50

        Shoes: $42.95

        Shorts: $19.50

2. In the problem above, about how much more did Norma pay for the pants than for the blouse? _____

3. About how much would 8 gallons of gasoline cost at $1.245 per gallon? _____

4. About how much did John's sales total after 3 days? _____

Day	Sales
1	$290.48
2	310.77
3	287.23

5. 4 friends sent out for pizza. If it cost $17.45 and they split it 4 ways, about how much did each person pay? _____

6. Rachel has been saving her change each day. About how much has she saved after 5 days? _____

Day	Change
1	$.42
2	.68
3	.25
4	.77
5	.14

7. Andy has $268.43 in his checking account. About how much would he have left if he wrote a check for $73.89? _____

8. Jack spent $65.89 on groceries for the week. About how much does he spend on food a day? _____

ANSWER KEY

Page 1: Learning About Estimation

1. Yes
2. Approximate numbers are reasonably close to the exact numbers, and they are easier and faster to work with.
3. About, nearly, almost, around, close to, etc.
4. c) $5
5. b) $400
6. a) $200

Page 2: Are the Answers Reasonable?

Estimates for questions 1–6 will vary depending on the experiences of each student. Some possible answers are below.

1. 8
2. 20
3. 5
4. 6
5. 25
6. $35

7. b) not reasonable
8. a) reasonable
9. b) not reasonable
10. a) reasonable
11. b) not reasonable·
12. b) not reasonable

Page 3: Decide When to Estimate

1. a) estimate
2. a) estimate
3. b) exact
4. a) estimate
5. b) exact
6. a) estimate
7. b) exact

Page 4: Estimating Metric Measurements

1. c) kilometers
2. a) meters
3. b) centimeters
4. a) centimeters
5. c) kilometers
6. c) meters

Page 5: Front-End Estimation

1. 9 tens = 90
2. 6 tens = 60
3. 7 tens = 70
4. 8 hundreds = 800
5. 17 hundreds = 1,700
6. 15 hundreds = 1,500

Page 6: Estimate Horizontal Forms

1. 14 tens or 140
2. 180
3. 180
4. 13 hundreds or 1,300
5. 1,600
6. 2,100
7. 14 thousands or 14,000
8. 21,000
9. 9,000

Page 7: Different Place Values for Front-End Digits

1. 6 hundreds = 600
2. 9 thousands = 9,000
3. 14 thousands = 14,000
4. 800
5. 14,000
6. 1,100
7. 200

Page 8: Applications for Front-End Estimation

1. a) 1,300
   b) 2,500

2. a) 500
   b) 800

3. a) 6,000
   b) 8,000
   c) 14,000

Page 9: Greatest Place Value

1. 500
2. 30
3. 7,000
4. 2,000
5. 30,000
6. 90
7. 20,000
8. 400
9. 3,000
10. 60,000
11. 100
12. 500
13. 2,000
14. 70,000

Page 10: Rounding to Estimate

1. 600 + 1,000 + 50
   1,650

2. 60 + 700 + 100
   860

3. 40 + 300 + 60
   400

4. 900 + 70 + 200
   1,170

5.
```
    100
    200
  + 100
    400
```

6.
```
   1,000
      40
  +  200
   1,240
```

7.
```
    500
     80
  +  80
    660
```

8.
```
     50
    300
  +  20
    370
```

Page 11: Rounding Larger Numbers to Add

1. 50,000 + 70,000 + 9,000
   129,000

2. 20,000 + 8,000 + 400
   28,400

3. 8,000 + 50,000 + 2,000
   60,000

4. 70,000 + 20,000
   90,000

5.   60,000
        6,000
   + 90,000
     156,000

6.   20,000
       3,000
   + 90,000
     113,000

7.   100,000
      40,000
   + 10,000
     150,000

8.   30,000
       5,000
   + 4,000
      39,000

Page 12: Grouping to 100

1. 200
2. 300
3. 200
4. 200
5. 300
6. 200
7. 300
8. 400

Page 13: Adjusting Front-End Estimation

1. a) 1,600 and about 200 more
   b) 1,800

2. a) 1,300 and about 200 more
   b) 1,500

3. a) 700 and about 200 more
   b) 900

4. a) 1,100 and about 300 more
   b) 1,400

Page 14: Adjusting Larger Numbers

1. a) 25,000 and about 3,000 more
   b) 28,000

2. a) 17,000 and about 3,000 more
   b) 20,000

3. a) 30,000 and about 2,000 more
   b) 32,000

4. a) 13,000 and about 3,000 more
   b) 16,000

Page 15: Apply Your Skills

Estimates will vary depending on which method the student uses (front-end, adjusted front-end, or rounding strategy).

1. a) $800–$1,000
   b) Answers will vary.
   c) $1,010

2. a) 11,000–12,000
   b) Answers will vary.
   c) 12,415

3. a) 270–300
   b) Answers will vary.
   c) 298

4. a) $400–$650
   b) Answers will vary.
   c) $622

5. a) 12,000–16,000
   b) Answers will vary.
   c) 16,175

6. a) 11,000–13,000
   b) Answers will vary.
   c) 13,103

Page 16: Estimating to Subtract

1. b) $15
2. a) $8,000
3. c) $5,500

Page 17: Rounding to Subtract

1.   70
   − 40
      30

2.   800
   − 200
     600

3.   5,000
   − 1,000
     4,000

4.   70,000
   − 50,000
     20,000

5. 200
6. 5,000
7. 60
8. 50,000

Page 18: Subtracting with Front-End Estimation

1. a) greater than 600
2. b) less than 400
3. b) less than 500
4. a) greater than 200
5. b) less than 200
6. a) greater than 200

Page 19: Adjusting Front-End Estimation

1. a) 40,000
   b) 39,000

3. a) 30,000
   b) 34,000

2. a) 30,000
   b) 33,000

4. a) 4,000
   b) 3,400

Page 20: Zero at Work

1. a) 6,000
   b) 5,900

4. a) 9,000
   b) 8,700

2. a) 5,000
   b) 5,400

5. a) 7,000
   b) 6,700

3. a) 1,000
   b) 1,300

6. a) 3,000
   b) 3,200

Page 21: Real-World Applications

Estimates will vary depending on which method the student uses to estimate. Exact amounts are given in parentheses.

1. $9,000–$11,000 ($9,410)
2. $100–$200 ($149)
3. 10,000–20,000 (14,520)
4. 50,000–60,000 (54,022)

Page 22: Make a Reasonable Guess

Estimates will vary. Exact amounts are given in parentheses.

1. a) 80,000–90,000 (85,766)
   b) 40,000–43,000 (42,015)

2. 140,000–153,000 (150,892)

3. a) 500–700 (563)
   b) 1,000–1,300 (1,131)
   c) 1,000–1,500 (1,358)

4. 2,000–4,200 (4,200)

5. a) $7,000–$7,300 ($7,183)
   b) $9,000–$10,000 ($9,448)
   c) $4,000–$5,000 ($4,839)

6. $16,000–$19,000 ($18,402)

Page 23: What Is Reasonable?

1. c) $50
2. a) $40
3. b) $18
4. a) $900
5. c) $500

Page 24: Estimating with Multiplication

1. 240
2. 6,300
3. 16,000
4. $80 \times 3 = 240$

5. $300 \times 5 = 1,500$
6. $9,000 \times 7 = 63,000$
7. $400 \times 2 = 800$
8. $6,000 \times 8 = 48,000$

Page 25: Rounding to Multiply

1. $90 \times 70 = 6,300$
2. $50 \times 30 = 1,500$
3. $80 \times 60 = 4,800$
4. $30 \times 10 = 300$

5. $40 \times 40 = 1,600$
6. $20 \times 40 = 800$
7. $60 \times 80 = 4,800$
8. $90 \times 20 = 1,800$

Page 26: Rounding to One-Digit Accuracy

1. $400 \times 20 = 8,000$
2. $600 \times 60 = 36,000$
3. $900 \times 20 = 18,000$
4. $400 \times 50 = 20,000$

5. $400 \times 400 = 160,000$
6. $100 \times 700 = 70,000$
7. $500 \times 600 = 300,000$
8. $300 \times 500 = 150,000$

Page 27: Is the Estimate High or Low?

1. About 6,300
   a) high

4. About 15,000
   c) difficult to tell

2. About 1,800
   b) low

5. About 24,000
   b) low

3. About 2,400
   c) difficult to tell

6. About 560,000
   a) high

Page 28: Estimating with a Map

Estimates will vary.

1. About 1,000 miles
2. About 1,400 miles
3. About 1,600 miles
4. About 2,100 miles
5. About 1,000 miles
6. About 2,700 miles

7. About 4 hours
8. About 6 hours
9. About 4 hours
10. About 8 hours
11. About 10 hours
12. About 4 hours

Page 29: Clustering

1. a) 500 miles
   b) 2,000 miles

2. a) $800
   b) $4,000

3. a) 5,000
   b) 20,000

4. a) 9,000
   b) 45,000

5. a) 10,000
   b) 60,000

Page 30: Does the Answer Make Sense?

1. b) $60
2. c) 50
3. b) $20
4. b) $200
5. a) 20

Page 31: Estimating with Division

1. a) 3 digits in the quotient
   b) hundreds

2. a) 2 digits in the quotient
   b) tens

3. a) 2 digits in the quotient
   b) tens

4. a) 4 digits in the quotient
   b) thousands

Page 32: Estimate the First Digit

1. Estimate: 700
2. Estimate: 90
3. Estimate: 600
4. Estimate: 4,000
5. Estimate: 40
6. Estimate: 600
7. Estimate: 2,000
8. Estimate: 200
9. Estimate: 10
10. Estimate: 80

Page 33: Estimates Hold the Key

1. Estimate: 600
2. Estimate: 30
3. Estimate: 200
4. Estimate: 70
5. Estimate: 2,000
6. Estimate: 7
7. Estimate: 60
8. Estimate: 7,000

Page 34: Compatible Numbers to Divide

Choices of compatible numbers and estimates may vary.

1. a) 4 and 20 are compatible.
   b) Estimate: 500

2. a) 7 and 42 are compatible.
   b) Estimate: 600

3. a) 6 and 30 are compatible.
   b) Estimate: 500

4. a) 9 and 54 are compatible.
   b) Estimate: 600

Page 35: Apply Your Skills

Estimates may vary.

1. About 900 boxes
2. About $300
3. About 30 miles per gallon
4. About 8 months
5. About $300

Page 36: Review of Whole Number Estimation

Estimates will vary. Exact amounts are given in parentheses.

1. About $400 ($407)
2. About $800 ($660)
3. About $100 ($103.60)
4. $1,000–$2,000 ($1,926)
5. About $1,800 ($1,770)
6. About 20 miles per gallon (20.86)
7. $19,000–$20,000 ($19,537)
8. $400–$500 ($478)

Page 37: Estimating with Decimals

1. Yes
2. No
3. Yes

Page 38: Front-End Estimation

1. 12, $12
2. 17, $17
3. 16, $16
4. 11, $110
5. 18, $180
6. 22, $220
7. 7, $700
8. 24, $2,400
9. 17, $1,700

Page 39: Rounding to Lead Digits

1. $8.00
2. $.50
3. $500
4. $30.00
5. $1.00
6. $5.00
7. $100.00
8. $ 30.00
9. $ .70
10. $500.00
11. $1.00
12. $50.00
13. $20.00
14. $600.00

Page 40: Add by Rounding to Lead Digits

1. $104.00
2. $49.00
3. $18.00
4. $990.00
5. $1,670.00
6. $89.00

Page 41: Estimating Dollar Amounts

1. Connect $.43 and $.62; $1.15
2. Connect $.48 and $.49; $1.75
3. Connect $.08 and $.88; $1.65
4. Connect $.39 and $.57; $1.05
5. Connect $.19 and $.79; $1.55
6. Connect $.45 and $.58; $1.20
7. Connect $.12 and $.90; $1.50
8. Connect $.51 and $.48; $1.29

Page 42: Adjusting Front-End Estimates

1. a) $11 and about $2.00 more
   b) Adjusted estimate: $13.00

2. a) $12 and about $2.00 more
   b) Adjusted estimate: $14.00

3. a) $9 and about $2.00 more
   b) Adjusted estimate: $11.00

4. a) $7 and about $2.00 more
   b) Adjusted estimate: $9.00

5. a) $14 and about $1.00 more
   b) Adjusted estimate: $15.00

6. a) $22 and about $2.00 more
   b) Adjusted estimate: $24.00

7. a) $7 and about $2.00 more
   b) Adjusted estimate: $9.00

8. a) $20 and about $3.00 more
   b) Adjusted estimate: $23.00

Page 43: Estimate Only Dollar Amounts

1. a) Front-end: $1,500
   b) Adjusted: $1,600
   c) Rounded: $1,500

2. a) Front-end: $140
   b) Adjusted: $160
   c) Rounded: $160

3. a) Front-end: $1,300
   b) Adjusted: $1,500
   c) Rounded: $1,500

4. a) Front-end: $1,100
   b) Adjusted: $1,200
   c) Rounded: $1,200

5. a) Front-end: $130
   b) Adjusted: $150
   c) Rounded: $160

6. a) Front-end: $1,600
   b) Adjusted: $1,700
   c) Rounded: $1,700

Page 44: Planning a Picnic

1. a) yes
2. Estimate: $24–$27
3. a) yes
4. About $5.00
5. About $30.00
6. a) yes
7. a) yes

Page 45: Is It Sensible?

1. b) $30
2. b) $2
3. c) $25
4. a) $15

Page 46: Rounding Decimals to Subtract

1.
$$\begin{array}{rr} \$7.23 & \$7 \\ -\ 3.58 & -\ 4 \end{array}$$
Rounded estimate: $3

2.
$$\begin{array}{rr} \$92.46 & \$90 \\ -\ 39.79 & -\ 40 \end{array}$$
Rounded estimate: $50

3.
$$\begin{array}{rr} \$72.45 & \$70 \\ -\ 28.98 & -\ 30 \end{array}$$
Rounded estimate: $40

4.
$$\begin{array}{rr} \$867.90 & \$900 \\ -\ 273.50 & -\ 300 \end{array}$$
Rounded estimate: $600

5. Rounded estimate: $40
6. Rounded estimate: $400
7. Rounded estimate: $4
8. Rounded estimate: $40

Page 47: Front-End Estimation and Subtraction

1. b) less than $6
2. a) greater than $3
3. b) less than $4
4. b) less than $1
5. a) greater than $2
6. a) greater than $7

Page 48: Adjusting Front-End Estimation

1. a) Front-end estimate: $20
   b) Adjusted estimate: $16

2. a) Front-end estimate: $40
   b) Adjusted estimate: $45

3. a) Front-end estimate: $200
   b) Adjusted estimate: $240

4. a) Front-end estimate: $40
   b) Adjusted estimate: $33

5. a) Front-end estimate: $300
   b) Adjusted estimate: $270

6. a) Front-end estimate: $40
   b) Adjusted estimate: $43

Page 49: Spring Specials

Estimates will vary. Exact amounts are given in parentheses.

1. a) Lawn mower: $80–$100 ($83.00)
   b) Running jacket: $10–$14 ($13.90)
   c) Running shorts: About $5 ($5.19)
   d) Outdoor grill: About $20 ($21.24)

2. $600–$700 ($639.34)

3. $15–$20 ($16.22)

4. $200–$230 ($228.20)

Page 50: Estimates That Make Sense

1. b) $2.60
2. c) $3.80
3. a) $4.25
4. b) $15.00
5. a) $11.00
6. c) $1.00
7. a) $85.00
8. b) $18.00
9. b) $50.00

Page 51: Shopping Trips

1. No
2. $10.00–$12.00
3. No
4. About $1.00
5. Yes
6. About $15.00

Page 52: Be Reasonable

1. b) $12
2. c) $120
3. a) $30
4. b) $5

Page 53: Round the Dollar Amounts

1. a) $8
   b) $3 × $8
   c) $24

2. a) $20
   b) 5 × $20
   c) $100

3. a) $3
   b) 6 × $3
   c) $18

4. a) $400
   b) 3 × $400
   c) $1,200

Page 54: Multiplying Mixed Decimals

1. $1.93 rounds to $2
   × 2.7 rounds to × 3
   Estimate: $6

2. $4.85 rounds to $5
   × 3.4 rounds to × 3
   Estimate: $15

3. $8.79 rounds to $9
   × 4.6 rounds to × 5
   Estimate: $45

4. $5.09 rounds to $5
   × 8 → × 8
   Estimate: $40

5. $1.69 rounds to $2
   × 2 → × 2
   Estimate: $4

6. $9.38 rounds to $9
   × 7 → × 7
   Estimate: $63

Page 55: Buying Fruits and Vegetables

1. 5 × $.50 = $2.50
   5 pounds of tomatoes will cost about $2.50.

2. 2 × $.80 = $1.60
   2 onion bunches will cost about $1.60.

3. 8 × $.60 = $4.80
   8 pounds of apples will cost about $4.80.

Page 56: Getting Closer

1. a) 5 × $4 = $20
   b) 5 × $.30 = $1.50
   c) $21.50

2. a) 3 × $1 = $3
   b) 3 × $.80 = $2.40
   c) $5.40

3. a) 2 × $3 = $6
   b) 2 × $.70 = $1.40
   c) $7.40

4. a) 4 × $8 = $32
   b) 4 × $.60 = $2.40
   c) $34.40

Page 57: Estimate the Costs

Estimates will vary. Exact amounts are given in parentheses.

1. a) $7.19   b) $14–$15 ($14.38)
2. a) $1.99   b) $3–$4 ($3.58)
3. a) $.89   b) $3.60–$4.00 ($3.56)
4. a) $6.73   b) $20–$22 ($21.54)
5. a) $.79   b) $3.00–$3.50 ($3.16)
6. a) $5.78   b) $11–$13 ($12.72)
7. a) $1.39   b) $2.00–$3.00 ($2.78)
8. a) $2.54   b) $7.00–$9.00 ($8.64)

Page 58: Clustering

1. a) 30 miles
   b) 150 miles

2. a) $10.00
   b) $40.00

3. a) $6,000
   b) $30,000

Page 59: Dividing to Estimate

1. **b)** $50
2. **a)** $1.25
3. **b)** $.60
4. **c)** $7.00

Page 60: Estimating to Divide

1. About $4.00
2. About $5.00
3. About $9.00
4. About $2.00
5. About $30.00
6. About $80.00
7. About $100.00
8. About $90.00

Page 61: Two-Digit Divisors

1. Estimate: 4.00
2. Estimate: .10
3. Estimate: 40.00
4. Estimate: 7.00
5. Estimate: 90.00
6. Estimate: 70.00

Page 62: Compatible Numbers to Divide

1. 8 and 72 are compatible.
   Estimate: $9

2. 5 and 25 are compatible.
   Estimate: $5

3. 6 and 30 are compatible.
   Estimate: $5

4. 7 and 28 are compatible.
   Estimate: $4

Page 63: Real-World Applications

Estimates will vary. Exact amounts are given in parentheses.

1. Estimate: About $20 ($19.83)
2. Estimate: About $90 ($91.65)
3. Estimate: About $100 ($108.99)
4. Estimate: About $8–$9 ($8.98)
5. Estimate: About $2–$3 ($2.92)
6. Estimate: About $3–$4 ($3.79)

Page 64: Estimate for Best Buy

1. Coffee: 39 ounces for $12.95
   Unit price estimate: $.30

   Coffee: 12 ounces for $4.85
   Unit price estimate: $.40

   The 39-ounce coffee has the lower unit price.

2. Detergent: 48 ounces for $3.59
   Unit price estimate: $.07

   Detergent: 28 ounces for $2.45
   Unit price estimate: $.08

   The 48-ounce detergent has the lower unit price.

Page 65: Practice Your Skills

Estimates will vary. Exact amounts are given in parentheses.

1. $8–$10 ($8.82)
2. $.80–$.90 ($.86)
3. About $9.00 ($9.36)
4. 180 miles (168.75)
5. <
6. >
7. >
8. <
9. >
10. <

Page 66: Ballpark Estimates

Estimates will vary. Exact amounts are given in parentheses.

1. $7,000–$8,000 ($7,778)
2. $4.50–$6.00 ($4.62)
3. $20,000–$22,000 ($22,090)
4. $13.00–$16.00 ($16.12)
5. About 80 miles (80.4)
6. No ($5.39)

Page 67: Reasonable Guess

Exact amounts are given in parentheses.

1. No ($5.49)
2. Yes ($9.15)
3. Yes ($9.46)
4. About $.90 ($.93)
5. About $4.80 ($4.72)
6. About $9 ($9.26)

Page 68: Number Sense

Exact amounts are given in parentheses. Note that calculators drop off end zeros in decimals: 12.10 will read 12.1.

1. **b)** No (742)
2. **a)** Yes (71.55)
3. **b)** No (28.22)
4. **b)** No (25,480)
5. **a)** Yes (12.1)
6. **b)** No (75.3)

Page 69: Review of Decimal Estimation

Estimates will vary. Exact amounts are given in parentheses.

1. $100–$123 ($125.90)
2. $10–$14 ($13.55)
3. $8–$10 ($9.96)
4. $700–$900 ($888.48)
5. $4–$5 ($4.36)
6. $2–$2.30 ($2.26)
7. $194–$230 ($194.54)
8. $9–$10 ($9.41)